A HALF-CENTURY OF AUTOMATA THEORY

A HALF-CENTURY OF AUTOMATA THEORY

Celebration and Inspiration

editors

A Salomaa
Turku Centre for Computer Science, Finland

D Wood
Hong Kong University of Science and Technology, China

S Yu
The University of Western Ontario, Canada

Published by

World Scientific Publishing Co. Pte. Ltd.

P O Box 128, Farrer Road, Singapore 912805

USA office: Suite 1B, 1060 Main Street, River Edge, NJ 07661

UK office: 57 Shelton Street, Covent Garden, London WC2H 9HE

Library of Congress Cataloging-in-Publication Data

A half-century of automata theory : celebration and inspiration / edited by Arto Salomaa, Derick Wood, Sheng Yu.

p. cm.

Includes bibliographical references.

ISBN 9810245904 (alk. paper)

1. Machine theory--Congresses. I. Salomaa, Arto, 1934– . II. Wood, Derick. III. Yu, Sheng.

QA267 .H333 2001

004--dc21 2001043518

British Library Cataloguing-in-Publication Data

A catalogue record for this book is available from the British Library.

Printed in Singapore by Mainland Press

PREFACE

In the past half century, automata theory has been established as one of the most important foundations of computer science, and its applications have spread to almost all areas of computer science. Research in automata theory and related areas has also reached a crucial point where researchers are searching for new directions.

To celebrate the achievements in automata theory in the past half century and to promote further research in the area, a celebratory symposium was organized, where eleven distinguished pioneers were invited to reminisce about the development of automata theory and to suggest possible future directions from their personal experiences and perspectives. The symposium, called "Half Century of Automata Theory" (HCAT), was held on Wednesday, July 26, 2000 at the University of Western Ontario, in London, Ontario, Canada,

The invited speakers included John Brzozowski, Sheila Greibach, Michael Harrison, Juris Hartmanis, John Hopcroft, Werner Kuich, Robert McNaughton, Maurice Nivat, Michael Rabin, Grzegorz Rozenberg (who was unable to attend due to sickness), and Arto Salomaa. Note that Juris Hartmanis, John Hopcroft, and Michael Rabin are ACM Turing Award winners.

The topics of their talks were ranging from general observations and predictions of the development of automata theory to specific topics which were considered to link the past to the future. In his talk at HCAT, John Hopcroft said: "Today there is another fundamental change taking place. We are in the situation right now which is very similar to what automata theory and theoretical computer science were in the 1960's. And I think in the next 20 years, it would be every bit as exciting". He also mentioned that "Automata theory may be poised for another period of significant advance, similar to that which occurred in the 1960's and 1970's".

Seven of the invited speakers contribute to this book. They are John Brzozowski, Juris Hartmanis, John Hopcroft, Werner Kuich, Robert McNaughton, Grzegorz Rozenberg, and Arto Salomaa. Their articles appear in the alphabetic order of their authors.

We wish to thank Ms. Chionh of the World Scientific for her cooperation in producing the volume. We wish also to thank Rachel Bevan, Rob Kitto, Clement Renard, Mark Sumner, and Geoff Wozniak for their efforts in writing the biographic information of the authors.

Arto Salomaa
Derick Wood
Sheng Yu

CONTENTS

HAZARD ALGEBRAS *
(EXTENDED ABSTRACT)

J. BRZOZOWSKI
Dept. of Computer Science, University of Waterloo
Waterloo, ON, Canada
brzozo@uwaterloo.ca

Z. ÉSIK
Dept. of Computer Science, University of Szeged
Szeged, Hungary
esik@inf.u-szeged.hu

We introduce algebras capable of representing, detecting, identifying, and counting static and dynamic hazard pulses on any wire in a gate circuit. These algebras also permit us to count the number of signal changes on any wire. This is of interest to logic designers for two reasons: each signal change consumes energy, and unnecessary multiple signal changes slow down the circuit operation. We describe efficient circuit simulation algorithms based on our algebras and illustrate them by several examples. Our method generalizes Eichelberger's ternary simulation and several other algebras designed for hazard detection.

1 Introduction

The problem of hazards, i.e., unwanted short pulses on the outputs of gates in logic circuits, is of great importance. In an asynchronous circuit a hazard pulse may cause an error in the circuit operation. Synchronous circuits are protected from such errors, since all actions are controlled by a common clock, and all combinational circuits stabilize before the clock pulse arises. However, an unwanted change in a signal increases the energy consumption in the circuit. From the energy point of view it is necessary not only to detect the presence of unwanted signal changes, but also to count them, in order to obtain an estimate of the energy consumption. Such unwanted changes also add to the computation time. In this paper we address the problem of counting hazards and signal changes in gate circuits.

One of the earliest simulation methods for hazard detection is ternary simulation. A two-pass ternary simulation method was introduced by Eichel-

*THIS RESEARCH WAS SUPPORTED BY GRANT NO. OGP0000871 FROM THE NATURAL SCIENCES AND ENGINEERING RESEARCH COUNCIL OF CANADA, AND GRANT NO. T30511 FROM THE NATIONAL FOUNDATION OF HUNGARY FOR SCIENTIFIC RESEARCH.

berger in 1965 [6], and later studied by others [3]. Ternary simulation is capable of detecting static hazards and oscillations, but does not detect dynamic hazards. A quinary algebra was proposed by Lewis in 1972 [10] for the detection of dynamic hazards. A survey of various simulation algebras for hazard detection was given in 1986 by Hayes [9]. We show that several of these algebras are special cases of the hazard algebra presented here.

The remainder of the paper is structured as follows. In Section 2 we discuss our model for representing transients in gate circuits. Using this model, in Section 3 we define Algebra C capable of counting an arbitrary number of signal changes on any wire in a gate circuit. To make the algebra more applicable, in Section 4 we modify it to Algebra C_k, where k is any positive integer; such an algebra is capable of counting and identifying up to $k-1$ signal changes. In Section 5 we describe circuit simulation algorithms based on Algebras C and C_k, and we illustrate our method by several examples. In Section 6 we extend our definitions to arbitrary Boolean functions. Complexity issues are then treated in Section 7, and Section 8 concludes the paper.

We omit all the proofs; they can be found in [1].

2 Transients

We use waveforms to represent changing binary signals. In particular, we are interested in studying transient phenomena in circuits. For this application we consider waveforms with a constant initial value, a transient period involving a finite number of changes, and a constant final value. Waveforms of this type are called *transients*.

Figure 1 gives four examples of transients. With each such transient we associate a binary word, i.e., a sequence of 0s and 1s, in a natural way. In this binary word a 0 (1) represents a maximal interval during which the signal has the value 0 (1). Such an interval is called a *0-interval* (*1-interval*). Of course, no timing information is represented by the binary word. This is to our advantage, however, since we assume that the changes can happen at any time and that the intervals between successive changes can vary arbitrarily.

We use $\vee$, $\wedge$, and $^-$ for the Boolean OR, AND, and NOT operations, respectively. For the present, we assume that our circuits are constructed with 2-input OR gates, 2-input AND gates and inverters. Given a transient at each input of a gate, we wish to find the longest possible transient at the output of that gate. For this reason, we need to assume that no pulses are lost, as might be the case if the delay of the gate were inertial. In this sense we are using the ideal delay model.

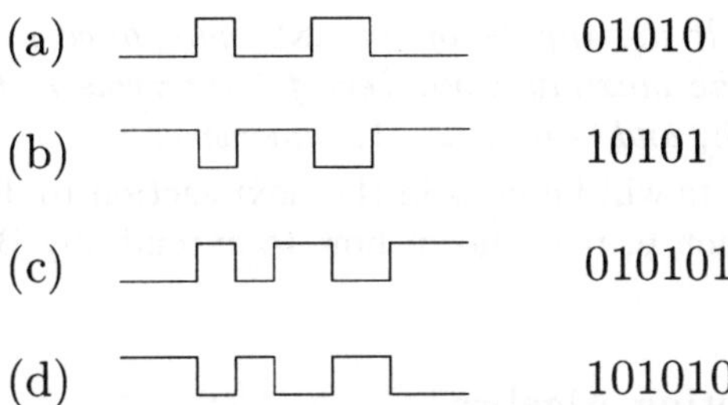

Figure 1. Transients: (a) constant 0 with static hazards; (b) constant 1 with static hazards; (c) change from 0 to 1 with dynamic hazards; (d) change from 1 to 0 with dynamic hazards.

We now examine how transients are processed by OR gates, AND gates, and inverters. The case of the inverter is the easiest one. If $t = a_1 \dots a_j$ is the binary word of a transient at the input of an inverter, then its output has the transient $\overline{t} = \overline{a_1} \dots \overline{a_j}$. For example, in Fig. 1, the first two transients are complementary, as are the last two.

For the OR and AND gates, we assume that the changes in each input signal can occur at arbitrary times. The following proposition permits us to find the largest number of changes possible at the output of a gate.

PROPOSITION 2.1 *If the inputs of an* OR *gate have m and n 0-intervals respectively, then the maximum number of 0-intervals in the output signal is 0 if $m = 0$ or $n = 0$, and is $m + n - 1$, otherwise.*

Example 1 *Figure 2 shows waveforms of two inputs X_1 and X_2 and output y of an* OR *gate. The input transients are* 010 *and* 1010, *and the output transient is* 101010. *Here, the inputs have two* 0*-intervals each, and the output has three* 0*-intervals, as predicted by Proposition 2.1.* □

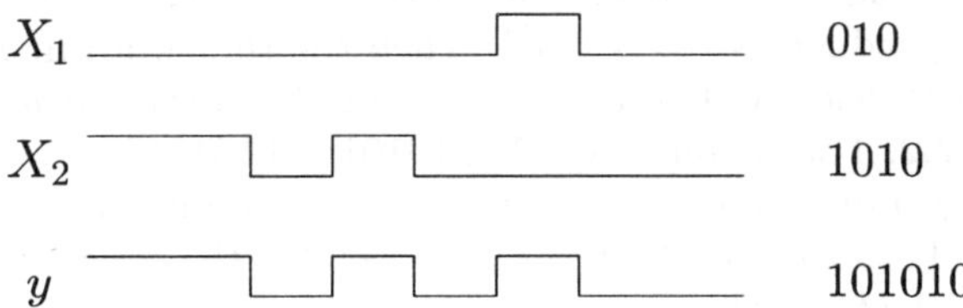

Figure 2. 0-intervals in OR GATE.

By an argument similar to that for Proposition 2.1, we obtain:

PROPOSITION 2.2 *If the inputs of an* AND *gate have m and n 1-intervals respectively, then the maximum number of 1-intervals in the output signal is 0 if $m = 0$ or $n = 0$, and is $m + n - 1$, otherwise.*

These two results will be used in the next section to define operations on transients. In Section 6, it is shown how to extend any Boolean function to transients.

3 Change-Counting Algebra

Let $T = 0(10)^* \cup 1(01)^* \cup 0(10)^*1 \cup 1(01)^*0$; this is the set of all nonempty words over 0 and 1, in which no two consecutive letters are the same. As explained above, elements of T are called transients.

We define the *(signal) change-counting algebra* $C = (T, \oplus, \otimes, ^-, 0, 1)$. For any $t \in T$ define $z(t)$ (z for zeros) and $u(t)$ (u for units) to be the number of 0s in t and the number of 1s in t, respectively. Let $\alpha(t)$ and $\omega(t)$ be the first and last letters of t, and let $l(t)$ denote the length of t. For example, if $t = 10101$, then $z(t) = 2$, $u(t) = 3$, $\alpha(t) = \omega(t) = 1$, and $l(t) = 5$.

Operations $\oplus$ and $\otimes$ are binary operations on T intended to represent the worst-case OR-ing and AND-ing of two transients at the inputs of a gate. They are defined as follows:

$$t \oplus 0 = 0 \oplus t = t, \quad t \oplus 1 = 1 \oplus t = 1,$$

for any $t \in T$. If w and w' are words in T of length > 1, and their *sum* is denoted by $t = w \oplus w'$, then t is that word in T that begins with $\alpha(w) \vee \alpha(w')$, ends with $\omega(w) \vee \omega(w')$, and has $z(t) = z(w) + z(w') - 1$, by Proposition 2.1. For example, $010 \oplus 1010 = 101010$, as illustrated in Fig. 2.

Next, define

$$t \otimes 1 = 1 \otimes t = t, \quad t \otimes 0 = 0 \otimes t = 0,$$

for any $t \in T$. Consider now the *product* of two words $w, w' \in T$ of length > 1, and denote this product by $t = w \otimes w'$. Then t is that word in T that begins with $\alpha(w) \wedge \alpha(w')$, ends with $\omega(w) \wedge \omega(w')$, and has $u(t) = u(w) + u(w') - 1$, by Proposition 2.2. For example, $0101 \otimes 10101 = 01010101$.

The *(quasi-)complement* $\overline{t}$ of a word $t \in T$ is obtained by complementing each letter in t. For example, $\overline{1010} = 0101$. Finally, the constants 0 and 1 are the words 0 and 1 of length 1.

A *commutative bisemigroup* is an algebra $C = (S, \oplus, \otimes)$, where S is a set, and $\oplus$ and $\otimes$ are associative and commutative binary operations on S, i.e., $(S, \oplus)$ and $(S, \otimes)$ are both commutative semigroups with the same underlying set. Thus a commutative bisemigroup satisfies equations L1, L2, L1′, L2′ in

Table 1, where the laws are listed in dual pairs. A commutative bisemigroup is *de Morgan* if it has two constants 0 and 1, and a unary operation $^-$ satisfying L3–L6, L3′, L4′, and L6′. All the laws of Table 1 are also satisfied by Boolean algebras, but several laws of Boolean algebras are not necessarily satisfied by de Morgan bisemigroups. Thus, de Morgan bisemigroups are generalizations of Boolean algebras.

Table 1. Laws of change-counting algebra.

L1	$x \oplus y = y \oplus x$	L1′	$x \otimes y = y \otimes x$
L2	$x \oplus (y \oplus z) = (x \oplus y) \oplus z$	L2′	$x \otimes (y \otimes z) = (x \otimes y) \otimes z$
L3	$x \oplus 1 = 1$	L3′	$x \otimes 0 = 0$
L4	$x \oplus 0 = x$	L4′	$x \otimes 1 = x$
L5	$\overline{\overline{x}} = x$		
L6	$\overline{x \oplus y} = \overline{x} \otimes \overline{y}$	L6′	$\overline{x \otimes y} = \overline{x} \oplus \overline{y}$

PROPOSITION 3.1 *The change-counting algebra* $C = (T, \oplus, \otimes, ^-, 0, 1)$, *is a commutative de Morgan bisemigroup, i.e., it satisfies the equations of Table 1.*

We note some further properties of the operations in C. For any two binary words t and t', t is a *prefix* of t' if there exists a (possibly empty) binary word t'' such that $t' = tt''$. The prefix relation is a partial order on the set of binary words, i.e., it is reflexive, antisymmetric and transitive. We use $\leq$ to denote the prefix order.

The prefix order restricted to T is represented by the inequalities below together with the reflexive and transitive laws:

$$0 \leq 01 \leq 010 \leq 0101 \leq 01010 \leq \ldots,$$

$$1 \leq 10 \leq 101 \leq 1010 \leq 10101 \leq \ldots.$$

A function $f(x_1, \ldots, x_n)$ defined on a partially ordered set, ordered by $\leq$, is *monotonic* if it is *monotonic* with respect to a partial order $\leq$ if and only if

$$x_1 \leq x'_1, \ldots, x_n \leq x'_n \text{ implies } f(x_1, \ldots, x_n) \leq f(x'_1, \ldots, x'_n).$$

For example, we know that $01 \otimes 10 = 010$. Since $\otimes$ is monotonic, as stated below, and since $01 \leq 0101$ and $10 \leq 10101$, we know that 010 is a prefix of the result $0101 \otimes 10101$. In fact, that result is 01010101.

PROPOSITION 3.2 *The* $\oplus, \otimes$ *and* $^-$ *operations are monotonic with respect to the prefix order.*

The next result shows that the length of a word t cannot be decreased by multiplying it by another word $s \neq 0$, or by adding another word $s \neq 1$.

COROLLARY 3.3 *Suppose that* $t, s \in T$. *If* $s \neq 1$, *then* $l(t) \leq l(t \oplus s)$. *If* $s \neq 0$, *then* $l(t) \leq l(t \otimes s)$.

For any word w, let w^{-1} denote the mirror image of w. It is clear that $w \in T$ iff $w^{-1} \in T$. Moreover,

$$\begin{aligned}(t \oplus s)^{-1} &= t^{-1} \oplus s^{-1} \\ (t \otimes s)^{-1} &= t^{-1} \otimes s^{-1} \\ \overline{t^{-1}} &= (\overline{t})^{-1},\end{aligned}$$

for all $t \in T$. Since also $(t^{-1})^{-1} = t$, we have:

PROPOSITION 3.4 *The function* $w \mapsto w^{-1}$, $w \in T$, *defines an automorphism of* C, *i.e., a one-to-one and onto mapping* $T \to T$ *which preserves the operations* $\oplus$, $\otimes$, $^{-}$, *and constants* 0 *and* 1.

It follows now that the operations also preserve the suffix order shown below.

$$0 \preceq 10 \preceq 010 \preceq 1010 \preceq 01010 \preceq \ldots,$$

$$1 \preceq 01 \preceq 101 \preceq 0101 \preceq 10101 \preceq \ldots.$$

4 Counting Changes to a Threshold

Since the underlying set T of Algebra C is infinite, an arbitrary number of changes can be counted. An alternative is to count only up to some threshold $k \geq 0$, and consider all transients with length k or more as equivalent.

Relation $\sim_k$ in algebra $C = (T, \oplus, \otimes, ^{-}, 0, 1)$ is defined as follows: For $t, s \in T$, $t \sim_k s$ if either $t = s$ or t and s are both of length $\geq k$.

PROPOSITION 4.1 *Relation* $\sim_k$ *is a congruence relation on* C, *by which we mean that it is an equivalence relation on* T *such that for all* $t, s, w \in T$, $t \sim_k s$ *implies* $(w \oplus t) \sim_k (w \oplus s)$, *and* $\overline{t} \sim_k \overline{s}$; *this then implies that* $(w \otimes t) \sim_k (w \otimes s)$ *whenever* $t \sim_k s$.

The equivalence classes of the quotient algebra $C_k = C/\sim_k$ are of two types. Each transient t with $l(t) < k$ is in a class by itself, and all the words of length $\geq k$ constitute a class. The operations on equivalence classes are as follows. The complement of the class containing t is the class containing $\overline{t}$. The sum (product) of the class containing t and the class containing t' is the class containing $t \oplus t'$ ($t \otimes t'$). Thus, the quotient algebra C_k is a commutative de Morgan bisemigroup with $2k - 1$ elements.

Example 2 *For* $k = 2$, *the* $\oplus$ *and* $\otimes$ *operations are shown in Table 2. These are the operations of the well known 3-element ternary algebra* [3]. □

Table 2. OR and AND operations for $k = 2$.

$\oplus$	0	Φ	1
0	0	Φ	1
Φ	Φ	Φ	1
1	1	1	1

$\otimes$	0	Φ	1
0	0	0	0
Φ	0	Φ	Φ
1	0	Φ	1

Example 3 *The* OR *and* AND *operations for the case $k = 3$ are shown in Table 3. This is the quinary algebra introduced in 1972 by Lewis* [10], *and studied also in* [4,5,9]. *Note that both binary operations are idempotent, i.e., laws L7 and L7′ hold; hence, the algebra is a bisemilattice* [4]. □

Table 3. OR and AND operations for $k = 3$.

$\oplus$	0	01	Φ	10	1
0	0	01	Φ	10	1
01	01	01	Φ	Φ	1
Φ	Φ	Φ	Φ	Φ	1
10	10	Φ	Φ	10	1
1	1	1	1	1	1

$\otimes$	0	01	Φ	10	1
0	0	0	0	0	0
01	0	01	Φ	Φ	01
Φ	0	Φ	Φ	Φ	Φ
10	0	Φ	Φ	10	10
1	0	01	Φ	10	1

We now discuss some further properties of Algebras C_k.

PROPOSITION 4.2 *Suppose that $k \geq 2$. Then $\sim_{k-1}$ is the smallest congruence relation strictly containing $\sim_k$.*

COROLLARY 4.3 *Each C_k with $k \geq 2$ is subdirectly irreducible.*

The last result means that C_k is not a subalgebra of a direct product of algebras with fewer elements. This implies, for example, that C_k cannot be expressed in any nontrivial way as an algebra of ordered triples [9], where each component of the triple is evaluated in its own algebra.

As we did for T, we define two partial orders on T_k; these are derived from the prefix and suffix orders. The class consisting of t is $\leq$ the class consisting of t' if and only if t is a prefix of t', and every class is $\leq \Phi$. The second partial order is similarly defined using suffixes. For example, the Hasse diagrams of the two orders on T_4 are shown in Fig. 3.

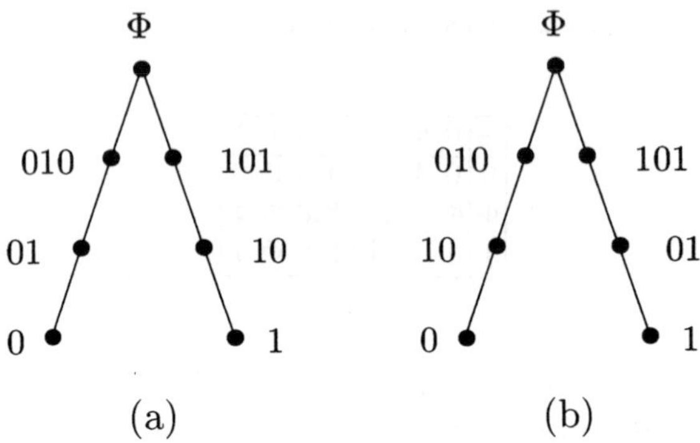

Figure 3. Partial orders on T_4: (a) $\leq$; (b) $\preceq$.

As was the case in C, the three operations $\oplus$, $\otimes$, and $^-$ are monotonic with respect to both partial orders. This property plays an important role in the simulation algorithms described in the next section.

5 Circuit Simulations

The binary analysis of a logic circuit with n gates may have as many as 2^n states. If one can be satisfied with partial information about the circuit behavior, then multivalued simulation is often an efficient method for finding that information.

For the detection of hazards, ternary algebra has been used since 1948 [8]. A two-pass ternary simulation method was introduced by Eichelberger in 1965 [6], and later studied by many authors [3]. The following example illustrates how static hazards are detected by ternary simulation.

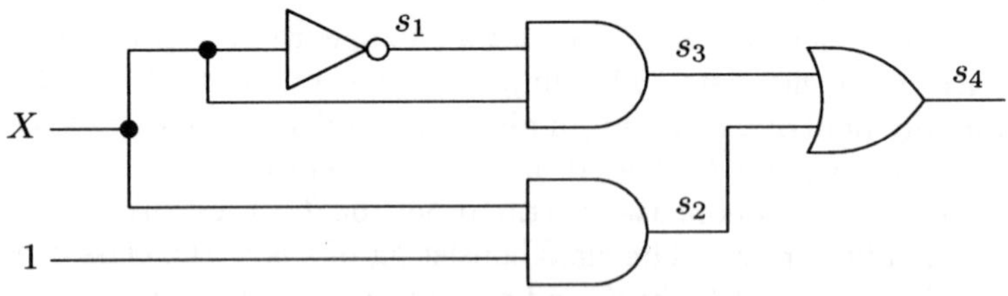

Figure 4. Circuit with hazards.

Table 4. Ternary simulation.

	X	s_1	s_2	s_3	s_4
initial state	0	1	0	0	0
	Φ	1	0	0	0
	Φ	Φ	Φ	Φ	0
result A	Φ	Φ	Φ	Φ	Φ
	1	Φ	Φ	Φ	Φ
	1	0	1	Φ	Φ
result B	1	0	1	0	1

Example 4 *Consider the circuit shown in Fig. 4. Let s_i denote the output of a gate, and S_i, the excitation of that gate. In the Boolean model we have the following excitation equations:*

$$S_1 = \overline{X},\ S_2 = 1 \wedge X,\ S_3 = X \wedge s_1,\ S_4 = s_2 \vee s_3.$$

Refer now to Table 4. The values of the input variable X and the state variables $s_1, \ldots, s_4$ are shown in rows as the simulation progresses. We begin in the initial state 01000, *which is stable. We wish to study the behavior of the circuit when the input changes from* 0 *to* 1 *and is kept constant at* 1.

Ternary simulation consists of two algorithms, A and B. In Algorithm A, the input changes to the uncertain or unknown value Φ. Instead of Boolean functions, we now use the ternary extensions of these functions as they are defined in Algebra C_2. Thus we use the excitation equations

$$S_1 = \overline{X},\ S_2 = 1 \otimes X,\ S_3 = X \otimes s_1,\ S_4 = s_2 \oplus s_3.$$

After X changes to Φ, Gates 1, 2, and 3, become unstable in the ternary model. In ternary simulation all unstable variables are changed at the same time. This results in the second row of Algorithm A. Now Gate 4 becomes unstable and changes, to yield the third row of Algorithm A.

In the case of C_2 the two partial orders $\leq$ and $\preceq$ coincide and are given by

$$0 \leq \Phi,\ 1 \leq \Phi,\ 0 \leq 0,\ 1 \leq 1,\ \Phi \leq \Phi.$$

This partial order is known as the uncertainty *partial order* [3].

In our example, the input became more uncertain by changing from 0 *to* Φ. *Since the gate operations preserve this order, the gate outputs can only become more uncertain. Thus the sequence of states produced by algorithm A is monotonically increasing, and the process terminates in at most n steps if there are n gates in the circuit. Intuitively, in Algorithm A we introduce uncertainty in the circuit inputs and we see how this uncertainty spreads throughout the circuit.*

In Algorithm B we start in the state produced by Algorithm A, but now we change the inputs to their final values from Φ*, thus reducing uncertainty. In our example,* X *becomes* 1*. We again use the ternary excitation functions to see whether the uncertainty will be removed from any gates. This time Algorithm B produces a monotonically decreasing sequence of states, which must terminate in no more than* n *steps. The final result is the vector* 10101*, showing that each gate reaches a binary value after the transient is over. Of course, this is the answer we expect since the circuit is feedback-free.*

It is clear from the circuit diagram that s_1 *changes from* 1 *to* 0 *and* s_2 *changes from* 0 *to* 1 *without any hazards. By Boolean analysis, it is easily verified that a static hazard may be present in* s_3*. It takes a little more effort to show that a dynamic hazard is present in* s_4*. Specifically, the variables may change in the following sequence* $s_3, s_4, s_1, s_3, s_4, s_2, s_4$*, if the gate delays have appropriate values* [3]*.*

This example illustrates that ternary simulation is capable of detecting static hazards. Gate s_3 *is* 0 *in the beginning and at the end of the transition, but it is* Φ *at the end of Algorithm A; this indicates a static hazard* [3]*. Our example also clearly shows that ternary simulation is not capable of detecting dynamic hazards. Gates* s_2 *and* s_4 *both change from* 0 *to* Φ *to* 1*, yet* s_2 *has no dynamic hazard, while* s_4 *has one.* □

In the examples that follow, we show how the accuracy of the simulation improves when we use Algebra C_k as k increases.

Example 5 *In Table 5 we repeat the simulation, this time using quinary extensions of gates as defined in Table 3. Instead to changing* X *to* Φ*, we change it to* 01 *in Algorithm A, since this is the transient we wish to study. From the result of Algorithm A we see that* s_1 *changes from* 1 *to* 0 *and* s_2 *changes from* 0 *to* 1*, both without hazards. The static hazard in* s_3 *is detected as before, as is the dynamic hazard in* s_4*. This example shows that quinary simulation is capable of detecting both static and dynamic hazards.* □

Example 6 *We repeat the simulation now using Algebra* C_4 *with seven values. Refer to Table 5. This time Algorithm A not only reveals that there is a static hazard in* s_3*, but also identifies it as* 010*. Note, however, that the dynamic hazard is still not identified.* □

Example 7 *We repeat the simulation using Algebra* C_5 *with nine values. Refer to Table 6. This time Algorithm A identifies the dynamic hazard as* 0101*.*

Table 5. Quinary and septenary simulation.

	X	s_1	s_2	s_3	s_4
initial state	0	1	0	0	0
	01	1	0	0	0
	01	10	01	01	0
	01	10	01	Φ	01
result A	01	10	01	Φ	Φ
	1	10	01	Φ	Φ
	1	0	1	10	Φ
result B	1	0	1	0	1

	X	s_1	s_2	s_3	s_4
initial state	0	1	0	0	0
	01	1	0	0	0
	01	10	01	01	0
	01	10	01	010	01
result A	01	10	01	010	Φ
	1	10	01	010	Φ
	1	0	1	10	Φ
result B	1	0	1	0	1

Table 6. Nonary simulation.

	X	s_1	s_2	s_3	s_4
initial state	0	1	0	0	0
	01	1	0	0	0
	01	10	01	01	0
	01	10	01	010	01
result A	01	10	01	010	0101
	1	10	01	010	0101
	1	0	1	10	0101
result B	1	0	1	0	1

Observe that the same table results if we simulate our circuit in Algebra C_k with any $k \geq 5$, or in Algebra C. Note also, that for $k \geq 4$ Algorithm B is no longer necessary, since the entire history of worst-case signal changes is recorded on each wire. □

Example 8 *Consider the circuit of Fig. 5. Algorithm A using Algebra C does not terminate, because there is a nontransient oscillation in the* OR *gate and the wire delay in the feedback loop. For more details see* [3]. *Simulation in Algebra C_k does, of course, terminate. In Table 7 we show the simulation with C_4. The nontransient oscillation is detected by the presence of Φ in the result of Algorithm B. This example also shows that multiple input changes can be handled by our simulation.* □

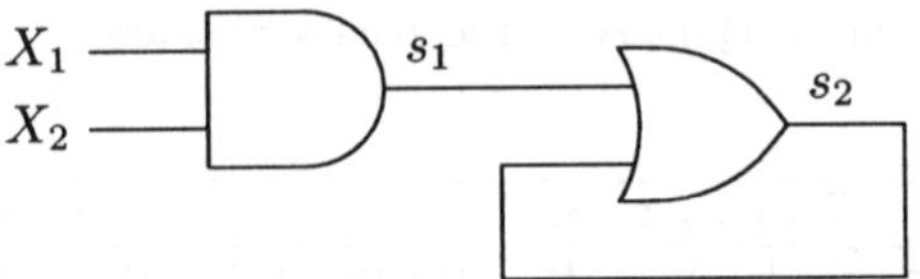

Figure 5. Circuit with nontransient oscillation.

Table 7. Simulation of circuit with nontransient oscillation.

	X_1	X_2	s_1	s_2
initial state	1	0	0	0
	10	01	0	0
	10	01	010	0
	10	01	010	010
result A	10	01	010	Φ
	0	1	010	Φ
result B	0	1	0	Φ

In a number of simulators [5,9], the signal values are ordered triples containing the initial, transient, and final values of a signal. Such algebras can also be described in our framework; for details see [1].

6 Extensions of Boolean Functions

In this section we show that simulation in Algebra C or C_k is not limited to 2-input OR and AND gates and inverters. Let $B = \{0, 1\}$. We now show how to extend any Boolean function $f : B^n \to B$ to a function $\hat{f} : T^n \to T$. We use the notation $[n]$ to mean $\{1, \ldots, n\}$.

Consider a simple example first. Suppose that a 2-input gate performs Boolean function f, and that the two inputs have transients 01 and 101, respectively. We want to find the maximum number of changes that can appear at the output of the gate, assuming that the input changes may occur at any time. Initially, the gate 'sees' the first letters of the two transients, i.e., 0 and 1; hence the output of the gate is $f(0, 1)$. One possibility is that the second transient occurs while the first input has the value 0. Then the gate would see the consecutive ordered pairs $(0, 1)$, $(0, 0)$, $(0, 1)$, and finally $(1, 1)$. Another possible order would be $(0, 1)$, $(0, 0)$, $(1, 0)$, and $(1, 1)$, and the third possible order would be $(0, 1)$, $(1, 1)$, $(1, 0)$, and $(1, 1)$. Note that we do not need to consider sequences like $(0,1),(1,0),(1,1)$, in which both inputs change in some steps, since each such sequence is a subsequence of one of the three sequences above. The corresponding output sequences in each of these cases would be $f(0,1), f(0,0), f(0,1), f(1,1)$, $f(0,1), f(0,0), f(1,0), f(1,1)$, and $f(0,1), f(1,1), f(1,0), f(1,1)$. If f is the OR function, we would have the sequences 1011, 1011, and 1111, respectively. The corresponding transients would be 101, 101, and 1. Since we are looking for the longest possible transient, we have 101 as the result of the operation $01 \oplus 101$, which, of course, agrees with our definition in Section 3. Similarly, if f is the AND function, we get the result 0101, and if it is the XOR function, we have 1010.

This example is generalized as follows. Suppose that $x = (x_1, \ldots, x_n)$, where $x_i \in T$, for each $i \in [n]$. Define the directed graph $D(x)$ to have as vertices n-tuples $y = (y_1, \ldots, y_n)$, where each y_i is a prefix of length > 0 of x_i, for each $i \in [n]$. There is an edge from vertex $y = (y_1, \ldots, y_n)$ to vertex $y' = (y'_1, \ldots, y'_n)$ iff y and y' differ in exactly one coordinate, say i, and $y'_i = y_i a$, where $a \in B$. For the example given above, we have the graph of Fig. 6.

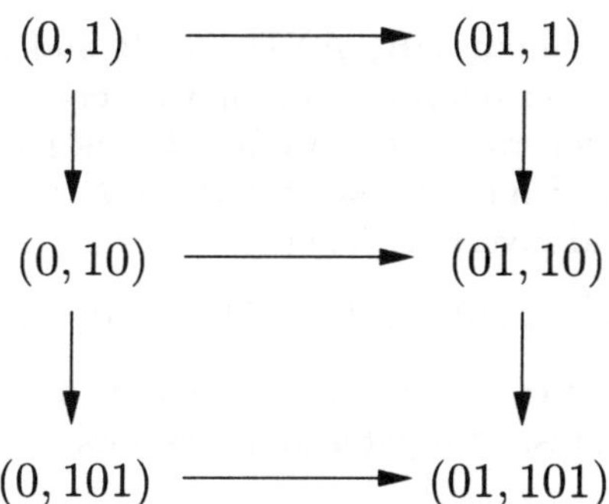

Figure 6. Graph $D(01, 101)$.

Let $x = b_1 \ldots b_n \in T$. Call a word y an *expansion* of x if y results from x by replacing each letter b_i, $i \in [n]$ by $b_i^{m_i}$, for some $m_i > 0$. Note that any nonempty word y over B is the expansion of a unique word $x \in T$. We call x the *contraction* of y.

Returning now to graph $D(x)$, we label each vertex $y = (y_1, \ldots, y_n)$ of $D(x)$ with the value $f(a_1, \ldots, a_n)$, where a_i is the last letter of y_i, i.e., $a_i = \omega(y_i)$, for each $i \in [n]$.

We define function $\hat{f}$ to be that function from T^n to T which, for any n-

tuple $(x_1, \ldots, x_n)$ of transients, produces the longest transient when $x_1, \ldots, x_n$ are applied to the inputs of a gate performing Boolean function f. The following proposition is now clear from the definition of $\hat{f}$ and the graph $D(x)$.

PROPOSITION 6.1 *The value of $\hat{f}(x_1, \ldots, x_n)$ is the contraction of the label sequence of those paths in $D(x)$ from $(\alpha(x_1), \ldots, \alpha(x_n))$ to $(x_1, \ldots, x_n)$ which have the largest number of alternations between* 0 *and* 1.

As an example, we now give a complete description of the extension of the n-input XOR function $f : B^n \to B$ to function $\hat{f} : T^n \to T$. For this function, no adjacent vertices have the same label. Hence, the length of the transient $\hat{f}(x_1, \ldots, x_n)$ is the length of any path in $D(x)$ from $(\alpha(x_1), \ldots, \alpha(x_n))$ to $(x_1, \ldots, x_n)$ (all such paths have the same length). One easily verifies that $y = \hat{f}(x_1, \ldots, x_n)$ is that word in T satisfying the conditions

$$\alpha(y) = f(\alpha(x_1), \ldots, \alpha(x_n))$$
$$\omega(y) = f(\omega(x_1), \ldots, \omega(x_n))$$
$$l(y) = 1 + \sum_{i=1}^{n}(l(x_i) - 1) = 1 - n + \sum_{i=1}^{n} l(x_i).$$

Extensions of the n-input OR, AND, NOR and NAND functions are similarly obtained; see [1] for details. Note that function composition does not preserve extensions in general. For a two input XOR gate with inputs 01 and 101 the output is 1010. Suppose now that the XOR gate is constructed with OR gates, AND gates and inverters. Then

$$(\overline{01} \otimes 101) \oplus (01 \otimes \overline{101}) = (10 \otimes 101) \oplus (01 \otimes 010) = 1010 \oplus 010 = 101010.$$

Thus, the hazard properties of the XOR function are quite different from those of the network of OR gates, AND gates and inverters.

PROPOSITION 6.2 *$\hat{f}$ is monotonic with respect to the prefix and suffix orders.*

It follows from the definition of $\hat{f}$ that the length of $\hat{f}(x_1, \ldots, x_n)$ is bounded by

$$1 + \sum_{i=1}^{n}(l(x_i) - 1) = 1 - n + \sum_{i=1}^{n} l(x_i).$$

When $n = 1$, this bound can be achieved for the identity function and function $^-$, and for $n > 1$, for the n-input XOR function.

PROPOSITION 6.3 *Suppose that f depends on each of its arguments and none of the x_i is a single letter. Then the length of $\hat{f}(x_1, \ldots, x_n)$ is at least the maximum of the lengths of the x_i.*

Proposition 6.4 *Let $k \geq 2$, and let $\sim_k$ be the equivalence relation on T, defined before. Then $\sim_k$ is a congruence in the sense that, for any Boolean function $f : B^n \to \{0,1\}$, and $x_1, \ldots, x_n, x'_1, \ldots, x'_n \in T$, if $x_1 \sim_k x'_1, \ldots, x_n \sim_k x'_n$, then $\hat{f}(x_1, \ldots, x_n) \sim_k \hat{f}(x'_1, \ldots, x'_n)$.*

Since $\sim_k$ is a congruence relation on T, it is meaningful to extend any Boolean function to T_k. It follows that the extension preserves the prefix and suffix order on T_k.

The XOR table for $k = 3$ is shown in Table 8.

Table 8. Operation of a XOR gate for $k = 3$.

XOR	0	01	Φ	10	1
0	0	01	Φ	10	1
01	0	Φ	Φ	Φ	10
Φ	Φ	Φ	Φ	Φ	Φ
10	10	Φ	Φ	Φ	01
1	1	10	Φ	01	1

7 Complexity Issues

In this section we give an estimate of the worst case performance of the simulation algorithms.

Using Algebra C, Algorithm A may not terminate unless the circuit is feedback-free. So suppose that we are given an m-input feedback-free circuit with n gates. We assume that each gate is an OR or AND gate, or an inverter, or more generally, a gate with a bounded number of inputs such that the extension of the binary gate function to T can be computed in linear time when each transient t is represented by the triple $(\alpha(t), z(t), \omega(t))$. Here, we assume that $z(t)$, the number of 0s in t is stored in binary.

Suppose the circuit is in a stable state and some of the inputs are supposed to change. If an input is to change from 0 to 1 (1 to 0), we set it to 01 (10). It takes $O(m)$ time to record these changes. Since the circuit contains no cycles, we may assume that the gates are given in topological order with corresponding state variables $s_1, \ldots, s_n$. Initially, each s_i has a binary value. Algorithm A then takes n steps. In step i, variable s_i is set to a new value according to its excitation. For example, when the ith gate is an OR gate which takes its inputs from the jth and kth gates, where $j, k < i$, then s_i is set to $s_j \oplus s_k$, the sum of the current values of s_j and s_k. The $\oplus$ operation is that of Algebra C. Using the triple representation of transients, the middle component of the new value of s_i is at most twice the maximum of the middle components of s_j and s_k, so that its binary representation is at most one longer than the maximum of the lengths of the middle components of s_j and s_k. We see that updating the value of s_i in the ith step takes $O(n)$ time. Thus, in n steps, we can set all n state variables to their final values in $O(n^2)$ time, showing that Algorithm A runs in $O(m + n^2)$ time. Since topological sort can be done within this time limit, the same bound applies if the gates are given in an arbitrary order.

By essentially the same algorithm, we can also compute the total number of signal changes in the entire circuit in $O(m + n^2)$ time. This total can be used as an indication of the worst-case energy consumption for a given input change of the circuit.

If the gates are unsorted, we need to update each state variable in each step. Since Algorithm A terminates in no more than n steps, and since in each step, each state variable can be updated in $O(n)$ time, this second method leads to the time bound of $O(m + n^3)$. However, if we use Algebra C_k, for a fixed k, then at each step, the value of any state variable can be stored in constant space and updated in constant time by a simple table look-up. In this case, Algorithm A terminates in $O(n)$ steps for all circuits, including those with feedback, since T_k is finite, and the subsequent values of each state variable form a monotonically increasing sequence with respect to the prefix order. Monotonicity follows from the fact, proved in [1] that the extension of any Boolean function preserves the prefix order on T_k. Thus, using Algebra C_k, Algorithm A runs in $O(m + n^2)$ time, even for circuits with feedback.

If we use Algorithm A with Algebra C for a feedback-free circuit, then Algorithm B becomes unnecessary. The last letter of the final value of each state variable gives the response of the circuit to the intended change in the input. Moreover, the final value of each state variable is the transient that describes all of the (unwanted) intermediate changes that can take place in worst case at the respective gate. The same holds if we use C_k for any circuit, which now may contain cycles, such that upon termination of Algorithm A, each state variable assumes a value other than Φ. However, if the final value of a state variable is Φ, Algorithm B does become necessary. It will stop in no more than n steps, since the subsequent values of each state variable now form a monotonically decreasing sequence with respect to the suffix order. The total time required by Algorithm B is $O(n^2)$.

8 Conclusions

We conclude the paper with a short summary of our results. Our main contribution is a general treatment of signal changes and hazards that encompasses the existing methods and permits a systematic study and comparison of these approaches. Some detailed properties of our method are highlighted below.

- **Hazards** We have presented a general theory of simulation of gate circuits for the purpose of hazard detection, identification, and counting.
- **Energy Estimation** The same simulation algorithms can be used to count the number of signal changes during a given input change of a circuit. This provides an estimate of the worst-case energy consumption of that input change.
- **Efficiency** If a circuit has m inputs and n gates, our simulation algorithms run in $O(m + n^2)$ time.
- **Accuracy** By choosing the value of the threshold k one can count signal changes and hazards to any degree of accuracy.
- **Feedback-Free Circuits**
  - In Algebra C, Algorithm A always terminates, and Algorithm B is not required.
  - In Algebra C_k, Algorithm A produces a result without Φs if k is sufficiently large. In that case, Algorithm B is not needed.
  - Simulation with (initial, middle, final) values terminates [1].
- **Circuits with Feedback**
  - In Algebra C, Algorithm A may not terminate.
  - Simulation in Algebra C_k always terminates.
  - Simulation with (initial, middle, final) values may not terminate; hence, it is not suitable for such circuits [1].
- **Multivalued Algebras** Many known algebras are included as special cases in our theory:
  - Ternary algebra is isomorphic to Algebra C_2.
  - The quinary algebra of Lewis [10] is isomorphic to Algebra C_3.
  - We can also represent the 6-valued algebra H_6 of Hayes [9] and the 8-valued algebra of Breuer and Harrison [2] and Fantauzzi [7,9] as quotients of C.

- **Decision Problems** [1]
  - For a given circuit in a stable initial state, a given input change, and an integer k, it is decidable in polynomial time whether there are k or more (unwanted) signal changes on the output of any given gate, or in the entire circuit, during that input change.
  - For a given circuit in a stable initial state and an integer k, it is decidable in nondeterministic polynomial time whether there exists an input change that would cause k or more (unwanted) signal changes on the output of any given gate, or in the entire circuit.

Characterizations of the simulation results described here in terms of the results of binary analysis, extensions to circuits started in unstable states, and other related results will be presented in a companion paper currently in preparation.

Acknowledgement

The authors thank Steve Nowick for his careful reading of our paper and for his constructive comments.

References

1. J. A. Brzozowski and Z. Ésik, "Hazard Algebras," Maveric Research Report 00-2, University of Waterloo, Waterloo, ON, http://maveric.uwaterloo.ca/publication.html, 27 pp., July 2000.
2. M. A. Breuer and L. Harrison, "Procedures for Eliminating Static and Dynamic Hazards in Test Generation," *IEEE Trans. on Computers*, vol. C-23, pp. 1069–1078, October 1974.
3. J. A. Brzozowski and C-J. H. Seger, *Asynchronous Circuits*, Springer-Verlag, 1995.
4. J. A. Brzozowski, "De Morgan Bisemilattices," *Proc. 30th Int. Symp. on Multiple-Valued Logic*, Portland, OR, pp. 173–178, IEEE Computer Society Press, Los Alamitos, CA, May 2000.
5. S. Chakraborty and D. L. Dill, "More Accurate Polynomial-Time Min-Max Simulation," *Proc. 3rd Int. Symp. on Advanced Research in Asynchronous Circuits and Systems*, Eindhoven, The Netherlands, pp. 112-123, IEEE Computer Society Press, Los Alamitos, CA, April 1997.
6. E. B. Eichelberger, "Hazard Detection in Combinational and Sequential Switching Circuits," *IBM J. Research and Development*, vol. 9, pp. 90–99, March 1965.
7. G. Fantauzzi, "An Algebraic Model for the Analysis of Logic Circuits," *IEEE Trans. on Computers*, vol. C-23, pp. 576–581, June 1974.

8. M. Goto, "Application of Three-Valued Logic to Construct the Theory of Relay Networks" (in Japanese), *Proceedings of the Joint Meeting of IEE, IECE, and I.of Illumination E. of Japan,* 1948.
9. J. P. Hayes, "Digital Simulation with Multiple Logic Values," *IEEE Trans. on Computer-Aided Design*, vol. CAD–5, no. 2, April 1986.
10. D. W. Lewis, *Hazard Detection by a Quinary Simulation of Logic Devices with Bounded Propagation Delays,* MSc Thesis, Electrical Engineering, Syracuse University, Syracuse, NY, January 1972.

UNDECIDABILITY AND INCOMPLETENESS RESULTS IN AUTOMATA THEORY

J. HARTMANIS

Cornell University
Ithaca, NY 14853
U.S.A.

Automata Theory is inextricably intertwined with undecidability and incompleteness results. This paper explores the beautiful panorama of undecidability and Gödel-like incompleteness results in automata theory that reveals their ubiquity and clarifies and illustrates how the severity of these results changes with the complexity of the problems and the computing power of the automata.

1 Introduction

Gödel's 1931 paper "Über formal unentscheidbare Sätze der Principia Mathematica und verwandter Systeme" [Go] sent shock waves through the mathematical community by showing that sufficiently rich mathematical theories can not be completely and consistently axiomatized (contrary to the belief of many mathematicians, including Hilbert and von Neumann). Any consistent axiomatization of a rich mathematical system has many true theorems that have no proofs in the system.

This signaled the end of a heroic but innocent mathematical era, epitomized by Hilbert's intellectual battle cry "Wir müssen wissen, wir werden wissen!", and the beginning of an intensive search for what is and is not effectively computable and formally provable. From this effort came the fundamental concept of a Turing machine and the Church-Turing thesis that the intuitive concept of effective computability is precisely captured by Turing machine computability or equivalently, by the class of recursive partial functions, and the subsequent development of recursive function theory.

A milestone in this development was Emil Post's 1944 address to the American Mathematical Society, "Recursively enumerable sets of positive integers and their decision problems" [Po]. This beautiful, easily readable paper clearly reveals the essence of recursively enumerable sets and their relation to Gödel's incompleteness results. It introduced the concept of reductions between problems, defined complete problems, and started the classification of recursively enumerable (r.e.) sets.

The development of computer science has been strongly influenced and aided by results and techniques from logic and recursive function theory. Computer science borrowed heavily from this rich intellectual arsenal and it owes a great debt of gratitude to these fields.

At the same time, computer science and, particularly Automata Theory, provide a panoramic view how undecidability and incompleteness results arise naturally in computer science research and how they limit what can be decided and proved about some very simple questions about automata.

Though Gödel's incompleteness result shocked the mathematical community, most mathematicians considered the result esoteric and did not expect that it would ever impact their work. This may still be true for much of mathematical research, but it certainly is not true for computer science. Computer science has to deal with many problems that do not have uniform decision procedures and it encounters incompleteness results in natural questions about computing systems.

In this paper, we illustrate how simply and elegantly one can derive a variety of undecidability and incompleteness results in automata theory. With the same techniques we derive non-recursive succinctness results between different representations of languages and give a simple proof of the existence of incomplete language in various complexity classes and show that the strongest incompleteness results can only hold for incomplete languages.

2 Basic Concepts and Π_2-Completeness

Let $M_1, M_2, \ldots; N_1, N_2, \ldots; D_1, D_2, \ldots$; and $G_1, G_2, \ldots$ be, respectively, the standard enumeration of Turing machines, (TM's), nondeterministic polynomial-time bounded TM's, deterministic polynomial-time bounded TM's and context-free grammars.

The Kleene Hierarchy provides an elegant classification of undecidable problems. The class Σ_1, consists of all r.e. sets and it can be characterized as the set of languages of the form

$$L = \{x \mid (\exists y)\ R\ [x, y]\},$$

where R is a recursive predicate. Π_1 is the corresponding class with a universal quantifier over a recursive predicate,

$$L = \{x \mid (\forall y) R[x, y]\}.$$

The Σ_2 class consists of all languages of the form

$$L = \{x \mid (\exists y)(\forall z)\ R\ [x, y, z]\}$$

and the Π_2 class of languages of the form

$$L = \{x \mid (\forall y)(\exists z)\ R\ [x, y, z]\ \},$$

with R a recursive predicate.

For example, the set of TM's that accept infinite sets, is a well known Π_2 set,

$$INF = \{M_i \mid (\forall n)(\exists y,t)[\mid y \mid \geq n \text{ and } M_i(y) \text{ accepts in } t \text{ steps }]\}.$$

The Σ_k and Π_k classes for $k > 2$ are defined correspondingly.

We say that a set A is *hard* for Π_k (Σ_k) iff for any B in Π_k (Σ_k) these exists a recursive function f such that

$$x \in B \leftrightarrow f(x) \in A\,.$$

A is *complete* for Π_k (Σ_k) if A is in Π_k (Σ_k) and A is hard for Π_k (Σ_k).

The following facts can be easily verified.

FACT 1. For all $k \geq 1$,

$$\Sigma_{k+1} \neq \Sigma_k \neq \Pi_k \neq \Pi_{k+1}.$$

FACT 2. The set INF is Π_2-complete.

Proof: Clearly,

$$INF = \{M_i \mid (\forall n)(\exists y,t)[\mid y \mid \geq n \text{ and } M_i(y) \text{ accepts in } t \text{ steps}]\}$$

is in Π_2.

Let A be a Π_2 set,

$$A = \{x \mid (\forall y)(\exists z)\ [R\ [x,y,z]\ = True\ ]\}.$$

Then A can be recursively reduced to INF as follows:

$$(\forall x)[f(x) = M_{\sigma(x)}]$$

where

$$M_{\sigma(x)}(y) \text{ accepts iff for all } u \leq y \text{ there exists a } z \text{ such that } R\,[x, u, z] = True.$$

Thus, $M_{\sigma(x)}$ is in INF iff x is in A.□

Since we are primarily interested in undecidability and incompleteness results in automata theory, we will illustrate these results for automata for which the membership problem is decidable. In particular, we consider context-free grammars and languages, because of their simplicity to illustrate these results. The following result about impossibility of naming all languages in Π_2-hard sets will yield many powerful undecidability and incompleteness results.
Let $A_1, A_2, \ldots$ be a standard enumeration of names for a family of automata.

We say that a set of automata accepting languages with property P,

$$\Delta = \{A_i \mid P\,[L(A_i)] = \mathit{True}\},$$

has an *r. e. list of names* if and only if there exists an r.e. list $A_{i_1}, A_{i_2}, \ldots$ that names all and only languages in Δ,

$$\{L(A_{i_j}) \mid j \geq 1\} = \{L(A_i) \mid P[L(A_i)] = \mathit{True}\}.$$

Let us now consider families of automata $A_1, A_2, \ldots$ for which the membership problem is solvable, i.e. x *in* $L(A_i)$ is recursively decidable. For example, the set of context-free grammars, the deterministic or nondeterministic polynomial-time Turing machines, etc. have a decidable membership problem.

Π_2-Lemma: If $\Delta = \{A_i \mid P\,[L\,(A_i)\,] = \mathit{True}\}$ is a Π_2-hard set then there is no r.e. list of names $A_{i_1}, A_{i_2}, \ldots$ such that

$$\{L(A_{i_j}) \mid j \geq 1\} = \{L(A_i) \mid A_i \text{ in } \Delta\}.$$

Proof : If there would be an r.e. list of names $A_{i_1}, A_{i_2}, \ldots$ then we could write Δ as a Σ_2 set:

$$\Delta = \{A_i \mid (\exists A_{i_j})(\forall x)[A_{i_j}(x) = A_i(x)]\}.$$

This is a contradiction to the assumption that Δ is a Π_2-hard set. Thus, no r.e. list of names can exist for Δ.□

3 Undecidability and Incompleteness Results

We say that a property on the r.e. sets is *non-trivial* if and only if there are r.e. sets for which the property is true and sets for which it is false.

Rice's Theorem: Any non-trivial property on r.e. sets is recursively undecidable for Turing machines.

The situation is quite different in automata theory. There are many decidable and undecidable problems of various difficulty and different varieties of incompleteness results. We will illustrate some of these results for context-free grammars and languages, which is one of the simplest classes of non-regular languages. We link context-free languages to Turing machine computations via *valid computations*. Recall that the valid computation of a Turing machine M_i on input x is the sequence of instantaneous descriptions of the computation of M_i on x, showing in each step the tape content, the tape square scanned by M_i and the state of M_i; starting with the start configuration and ending in the accept configuration. For details see [HU]. The set of all valid computations of M_i is denoted by $VAL(M_i)$.

It can easily be seen that a nondeterministic push-down automaton can accept the complement of $VAL(M_i)$, $\overline{VAL(M_i)}$. The push-down automaton can nondeterministically guess where an error could occur and then (using its stack as distance measure) compare the corresponding symbols in the following instantaneous description. If an error is found the string is not in $VAL(M_i)$ and it is accepted.

From the above it follows that for each M_i we can effectively construct $G_{\sigma(i)}$ such that

$$L(G_{\sigma(i)}) = \Sigma^* - VAL(M_i) = \overline{VAL(M_i)}\,.$$

We observe that

$$L(M_i) = \emptyset \text{ iff } VAL(M_i) = \emptyset \text{ iff } \overline{VAL(M_i)} = \Sigma^*.$$

Since $\overline{VAL(M_i)} = L(G_{\sigma(i)})$ and it is recursively undecidable if $L(M_i) = \emptyset$, we see that for context-free grammars it is recursively undecidable if $L(G_i) = \Sigma^*$.

Recall that it is recursively decidable if $L(G_i) = \emptyset$ and if $L(G_i)$ is infinite. On the other hand, for deterministic polynomial-time Turing machines, $D_1, D_2, \ldots$, it is recursively undecidable if $L(D_i) = \emptyset$ and if $L(D_i)$ is infinite.

Every undecidability result implies a Gödel-like incompleteness result. Let us consider a sound axiomatizable formula system F, that is, F proves only true theorems and the set of provable theorems is r.e.

The above observation that for context-free grammars the problem $L(G_i) = \Sigma^*$ is recursively undecidable, implies the following incompleteness result.

Theorem: For any sound axiomatizable F there exists a G_i with $L(G_i) = \Sigma^*$ and there is no proof in F that "$L(G_i) = \Sigma^*$."

Proof: We know that the set

$$\{G_i \mid L(G_i) \neq \Sigma^*\}$$

is recursively enumerable. If for each $G_i, L(G_i) = \Sigma^*$, there would be a proof in F of this fact, then the set

$$\{G_i \mid L(G_i) = \Sigma^*\}$$

would be r.e. and therefore recursive. This is a contradiction and we conclude that there is no proof in F that "$L(G_i) = \Sigma^*$." □

4 Representation Independent Incompleteness Results

The previous incompleteness result is not very interesting in that it asserts that for some context-free grammar G_i with $L(G_i) = \Sigma^*$, there is no proof in F that "$L(G_i) = \Sigma^*$." In any sufficiently strong formal system there are simple grammars for which we can prove in F that "$L(G_i) = \Sigma^*$." Thus the above result exploits a bad grammar for Σ^* that F cannot "understand."

In short, the above incompleteness result depends on specific names for a simple language and does not depend on the complexity of the language itself.

Next we show how the Π_2-Lemma yields representation independent incompleteness results. These results are true statements about languages not provable for any representation of these languages. These *machine independent incompleteness results* are based on the complexity of the property under consideration independent of how the languages are represented.

We recall that for Turing machines the set

$$\{M_i \mid |L(M_i)| = \infty\}$$

is Π_2-complete. Since

$L(M_i) = \infty$ iff $| VAL(M_i) | = \infty$ iff $VAL(M_i)$ is not regular iff $\overline{VAL(M_i)} = L(G_{\sigma(i)})$ is not regular, we can reduce a Π_2-complete set to the Π_2 set

$$\{G_i \mid L(G_i) \text{ is not regular}\}$$

which shows that it is a Π_2-complete set.

From this we get a representation independent incompleteness result.

Theorem: For any sound axiomatizable system F there exist a non-regular context-free language $L(G_i)$ such that for no grammar $G_j, L(G_j) = L(G_i)$, is there a proof in F that "$L(G_j)$ is not regular."

Proof: Assume that for every non-regular context-free language L there is some context-free grammar $G_i, L(G_i) = L$, and a proof in F that "$L(G_i)$ is not regular." Then we would have an r.e. list of names for the Π_2-complete set

$$\{G_i \mid L(G_i) \text{ is not regular}\}$$

which is impossible by the Π_2-Lemma. Thus, there must exists a non-regular context-free language for which the non-regularity cannot be proved in F for any representation of this language. □

By the same technique we can show several other results about representation independent incompleteness results.

Corollary: If F is a sound axiomatizable system and $P \neq NP$, then there exists an L in $NP - P$ such that for no $N_i, L(N_i) = L$, is there a proof in F that "$L(N_i)$ is not in P."

Corollary: If F is a sound axiomatizable system then there is a context-free grammar G such that $\Sigma^* - L(G)$ is not a context-free language and for no G_i for $L(G)$ is there a proof in F that

$$"\Sigma^* - L(G_i) \text{ is not a cfl."}$$

We can also obtain incompleteness results about the P versus NP problem. Let E_i denote deterministic exponential space machines. Since we can show that

$$\{E_i \mid NP^{L(E_i)} \neq P^{L(E_i)}\}$$

is Π_2-complete, we have the next result.

Theorem: For any sound axiomatizable F there exists an $E_{i_\circ}$ such that

$$NP^{L(E_{i_\circ})} \neq P^{L(E_{i_\circ})}$$

and for no E_j equivalent to $E_{i_\circ}$ is there a proof in F for

$$"NP^{L(E_j)} \neq P^{L(E_j)}" \text{ or } "NP^{L(E_j)} = P^{L(E_j)}"$$

It is interesting to contrast the above result with another related result [HH].

Theorem: For any sound axiomatizable F there is an effectively constructible Turing machine $M_{i_\circ}$ such that there is no proof in F for

$$"NP^{L(M_{i_\circ})} \neq P^{L(M_{i_\circ})}" \text{ or } "NP^{L(M_{i_\circ})} = P^{L(M_{i_\circ})}"$$

and $L(M_{i_\circ}) = \emptyset$.

5 Incomplete Languages

In 1975 Ladner [La] showed that if $P \neq NP$ then there exist incomplete languages in $NP - P$. We derive this result by observing that if P is not NP

then

$$\Delta = \{N_i \mid L(N_i) \text{ in } NP - P\}$$

is a Π_2-complete set. Therefore Δ cannot have an r.e. list of names. On the other hand, it is not hard to show that the complete language in NP have an r.e. list of names [LLR]. This implies that if $P \neq NP$ then there exist incomplete languages in $NP - P$.

Furthermore, we now observe that the strong, representation independent incompleteness results, can only hold for incomplete languages in $NP - P$.

We assume that our sound and axiomatizable proof system F is sufficiently strong to prove correct simple automata constructions, such as the construction of an r.e. list of names for the complete languages in NP.

Theorem: Let F be a sound axiomatizable system that proves "$P \neq NP$." Then there exists A in $NP - P$ for which there is no proof in F for any $N_i, L(N_i) = A$, that "$L(N_i)$ is not in P." Furthermore, A must be an incomplete language.

Proof: Since F proves "$P \neq NP$" we know that $P \neq NP$ and therefore the set

$$\Delta = \{N_i \mid L(N_i) \ in \ NP - P\}$$

is Π_2-complete. If for every A in $NP - P$ there would be a proof in F for some N_i which accepts A that "$L(N_i)$ is not in $NP - P$," then there would be an r.e. list of names for Δ, contradicting that Δ is Π_2-complete. Furthermore, since for every complete language L, there is some N_i that accepts L and F proves "$L(N_i)$ is not in P", we see that the languages in the above proof must be incomplete languages in $NP - P$.□

It should be emphasized that there are many N_i that accept complete sets in $NP - P$ for which there is no proof in F that "$L(N_i)$ is not in P." But for each such N_i there is some other N_j accepting the same language for which there is a proof in F that "$L(N_j)$ is not in P." Only for the incomplete languages A in $NP - P$ can we get the stronger incompleteness result that for no N_i accepting A is there a proof in F that "$L(N_i)$ is not in P."

6 Minimal Automata and Incompleteness

It is well known that deterministic finite automata can be effectively minimized and therefore the set of minimal finite automata is recursive.

To define minimal context-free grammars we impose a natural linear order ($\leq$) on the set of these grammars. We fix a finite alphabet Σ to describe the grammars and then order them by the length of description and lexicographically order the grammars of the same length.

Theorem: The set of minimal context-free grammars

$$\{G_i \mid G_j < G_i \text{ implies } L(G_i) \neq L(G_j)\}$$

is recursively enumerable but not recursive.

Proof: G_i is minimal if and only if it generates a language different from the language generated by the finite number of smaller grammars than G_i. If this is the case, it can be detected and therefore the above set is r.e..

To see that the set of minimal grammars is not recursive, observe that if it would be recursive then for each G_i we could effectively find the unique minimal grammar equivalent to G_i. But then we could solve the equivalence problem for context-free languages, which is not solvable. □

We conjecture that the set of minimal context-free grammars (and similarly the set of minimal linearly-bounded automata, etc.) is not a many-to-one complete r.e. set, but have not been able to find a proof.

Finally, it can easily be shown by the use of Recursion Theorem or by a direct proof that the set of minimal Turing machines is immune (first shown by M. Blum). That is, any r.e. subset of minimal Turing machines is finite. From this we get a very strong incompleteness result.

Theorem: Any sound axiomatizable formal system F can prove only a finite number of Turing machines to be minimal.

Actually, the size of the formal system F (described in a fixed alphabet as a TM that accepts the set of provable theorems) determines an upper bound of how many Turing machines F can prove minimal.

Corollary: There is a positive constant $k_\circ$ such that for any sound axiomatizable F there is no proof of "M_i is minimal" in F if $\mid M_i \mid \geq \mid F \mid + k_\circ$.

7 Succinctness Results

We conclude by exploiting the Π_2-completeness of

$$\{G_i \mid L(G_i) \text{ is not regular}\}$$

to derive non-recursive succinctness results about representation of regular languages, by finite automata and context-free grammars, respectively.

Theorem [MF]: There is no recursive bound F: $N \rightarrow N$ such that for any context-free grammar G_i generating a regular set there exist an equivalent finite automaton $A_j, L(G_i) = L(A_j)$, and

$$F(|G_j|) > |A_i|.$$

Proof: If such recursive F exists then we can recursively enumerate the set

$$\{G_i \mid L(G_i) \text{ not regular}\}$$

by checking if $L(G_i)$ differs from the finite number of regular languages $L(A_j), |A_j| \leq F(|G_i|)$.

But this contradicts the Π_2-completeness of

$$\{G_i \mid L(G_i) \text{ not regular}\}.$$

□

A similar proof works if P does not equal NP for the Π_2-complete set

$$\{N_i | L(N_i) \text{ not in } P\}$$

and shows that there is no recursive succinctness bound for the representation of sets of P by deterministic and nondeterministic polynomial-time machines, respectively.

Many other results of this type follow once the Π_2-completeness of the respective sets is established.

8 Conclusion

We believe that Hilbert and Gödel would indeed be surprised, for different reasons, how undecidability and incompleteness results arise naturally in automata theory. Furthermore, it is interesting to observe that the provable nonexistence of uniform algorithmic solutions of practical problems in computer science is wide spread and directly affects the work of computer scientists.

Acknowledgment

Some of the ideas and concepts of this paper appear in [Ha].

References

[Go] Gödel, K. "Über formal unentscheidbare Sätze der Principia Mathematica und verwandter Systeme", Monatsheft für Mathematik und Physik, 38:173-198, 1931.

[Ha] Hartmanis, J. "On the importance of being Π_2-hard", Bulletin of the European Assoc. of Theoretical Computer Science, (EATCS), vol 37:117-127, Feb. 1989.

[HH] Hartmanis, J and J.E. Hopcroft, "Independence results in computer science", ACM SIGACT News, Vol 8, No. 4: 13-24, Oct.-Dec. 1976.

[HU] Hopcroft, J.E. and J.D. Ullman. *Introduction to Automata Theory, Languages, and Computation.* Addison-Wesley, Reading, MA, 1979.

[La] Ladner, R.E. "On the Structure of Polynomial Time Reducibility", J. ACM 22:155-171, 1975.

[LLR] Landweber, L.H., R.J. Lipton and E.L. Robertson, "On the structure of sets in NP and other complexity classes", Theoretical Computer Science 15: 181-200, 1981.

[MF] Meyer, A.R. and M.J. Fischer, "Economy of Descriptions by Automata, Grammars, and Formal Systems", 12th IEEE Symp. on Switching and Automata Theory: 188-191, 1971.

[Po] Post, E. "Recursively Enumerable Sets of Positive Integers and Their Decision Problems", Bulletin AMS 50:284-316, 1944.

AUTOMATA THEORY: ITS PAST AND FUTURE

JOHN HOPCROFT
Cornell University
Ithaca, NY 14853, USA

This paper briefly reviews key advances in automata theory since 1960. It focuses on why certain lines of inquiry were followed, why others were not, and why some works were deemed more valuable to the discipline than others. I also compare the field today to what it was in 1960, with an eye toward areas in which the biggest breakthroughs are likely to occur in the future. Theoretical computer science may be poised for another period of significant advance, similar to that which occurred in the 1960's and 70's.

1 Introduction

This paper briefly reviews key advances in automata theory over the past fifty years, with an eye toward why certain directions were taken, and why people worked on certain problems. First, we will survey the major themes which drove early research in automata theory and influenced the evolution of courses in theoretical computer science. Then we will examine how the broader field of computing has changed, and how these changes are impacting the direction that theoretical computer science will have to take in order to support the computing activities of the current age. Theoretical computer science may be poised for another period of significant advance, similar to that which occurred in the 1960's and 70's.

2 Beginnings of Automata Theory

This talk begins around the middle of the twentieth century. The electrical engineers of the late 1950's who were working on switching circuit theory and logical design are among the forerunners of the field of automata theory. At that time, computer construction was the focus. These engineers were educating the students who were going to go out and build computers and design circuits.

2.1 1960: Formal Languages

Around 1960, people recognized that computing was going to change the world. Funding agencies supported computing, universities started to create computer science departments, and some of the very best students were moving into computer science. People also began to realize that it was not the construction of computers that was important, but also how we were going to program them. How would we write compilers? Operating systems? This is when the discipline started to work on finite state automata, formal languages, and computability.

Consequently, faculty began to move from teaching switching theory to students to teaching more theoretical material. However, around that time there were very few theory textbooks. To teach a course in automata theory, one likely developed the course around the book *Automata Theory: Selected Papers* by E. F. Moore, and the following three key papers:

(1) *Regular expressions and state graphs for automata*, by R. McNaughton, H. Yamada;

(2) *Finite Automata and their decision problems*, by M.O. Rabin, Dana Scott;

(3) *The reduction of two-way automata to one-way automata*, by J.C. Shepherdson.

Regular expressions, the subject of the first paper, had their beginnings in interest surrounding neurons and firing sequences. As we know, it turned out that this notation happened to say a lot about other things, as well.

In the second paper, a nondeterministic model of finite automata was introduced. It was shown that this model was equivalent to the deterministic one. The fact that two models which were so different gave rise to the same sets, along with the fact that both were equivalent to the regular expression notation, indicated that these models were fundamental. This paper was very influential in shaping automata theory, and the concept of nondeterminism played an influential role in later events.

The third paper, not quite as influential, showed that finite automata with the ability to scan their input in both directions still gave rise to the same sets. The invariance of the model to changes in the definition suggested that the model was fundamental and that a rich set of theorems was likely to be provable from it. Of course, for academics these were all good reasons for continuing to study automata theory. But when we taught classes we gave other reasons, such as the fact that finite state devices model certain kinds of controls, and are used for pattern recognition in text and the lexical portion of parsing.

In addition to regular expressions and finite automata, a theory course also contained material on context-free grammars. The context-free grammar became important when it was found to be equivalent to the nondeterministic pushdown automata model. It was then realized that this was a natural notation, and interesting things could probably be proven about it. Context-free grammars captured (or we claimed that it captured) the syntax of programming languages, which had become an important topic. At that time, writing compilers was a difficult task. There was a belief that at least the construction of the parsing portion of the compiler could be automated with the aid of context-free grammars, making it easier to teach students how to write a compiler.

The deterministic pushdown automata model was important for course content, since computer memory was limited, forcing programs to be parsed sequentially. However, it was observed that converting a context-free grammar representation for a language that could be recognized by a deterministic pushdown automata model to a deterministic context-free grammar resulted in a size explosion. That got people interested in questions about representations and size of representations.

The remaining topics that were put into theory courses were computability and complexity. The paper *On the computational complexity of algorithms* by Hartmanis and Stearns was likely used for teaching complexity theory. In that paper, it was shown that small increases in time or space would allow new computations.

One of the reasons that undecidability was put into the course was probably just a fortuitous accident. The people who designed the grammar for Algol 60 happened to put together a grammar which was ambiguous in one particular case (see Figure 1). This raised the question "can we avoid difficulties like this by writing an algorithm that takes a context-free grammar and determines whether or not it is ambiguous?". The answer was no. The problem was recursively undecidable. This led to a realization in the computer science community that there were real problems, not just mathematical problems, that were undecidable and maybe we ought to teach a little bit about undecidability.

Another idea that drove early research was the attempt to prove lower bounds. One technique developed to prove a lower bound on a single-tape Turing machine was the notion of crossing-sequences. For example, it was shown that to recognize the language ww^r took at least n^2 steps on a single-tape Turing machine. Somehow, the TM had to go back and forth, maybe not in a regular pattern, but there was some region of the input that had to be crossed at least $n/2$ times. Therefore the length of the crossing-sequences

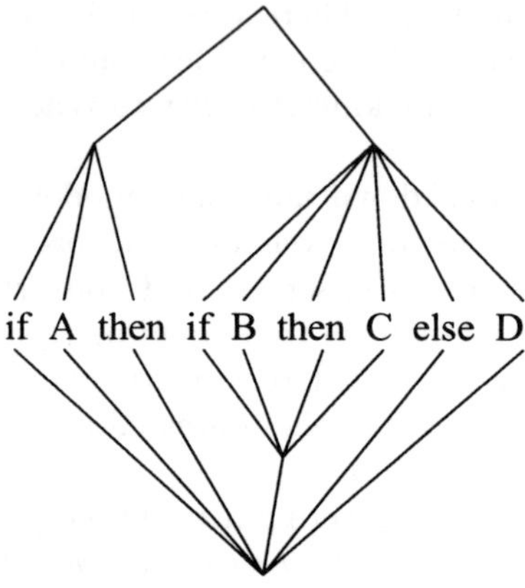

Figure 1. Ambiguity in Algol60

in that region was $n/2$, giving an n^2 lower bound.

Researchers then tried to develop other lower bounds, and they noticed there was something interesting here. One way to get lower bounds is by diagonalization (for example you can diagonalize over all problems that can be solved in time n^3 to show something can't be done in time n^3). But that doesn't usually give a problem that arises naturally. Therefore, people tried what were called "information transfer arguments", which is what the idea of crossing-sequences is: a certain amount of information over "here" has to be transferred over "there", and to transfer that amount of information takes a certain amount of time or space, and so on.

What was interesting, was we were never able to push these lower bounds beyond the trivial. This is why there were many papers published which, looking back, don't seem worthy of publication. For example, in one paper it was shown for a circuit complexity problem that something like $3.5n$ gates were required. The author was unable to show that the circuit required more than a linear number of gates. All the author really managed to do was to show the constant out in front was greater than 2. This was a common problem, and led to another theme.

To summarize, there were five important themes which drove early research in theoretical computer science: invariant models, equivalent notations, size of representations, lower bounds, and information theory/diagonalization.

2.2 1970: Analysis of Algorithms

Around 1970, there was a change of focus from compilers and the programming languages used to express algorithms to the algorithms themselves. Most computer science students were not going to become theoretical computer scientists. Thus, the analysis of algorithms was added to the computer science curriculum to educate students who would ultimately be employed writing computer programs for various applications.

What were important topics in algorithms? When comparing computer programs, we didn't just want to say that one program is better than another because one runs in 17 seconds and the other runs in 23 seconds! In the literature at that time, when a new algorithm was published, some running times on some data would always be included. When the next person published his algorithm, he would run it on the data that the previous person had, and would invariably have faster running times. But it was never clear how much of that was due to the algorithm, how much was due to an increase in computer speed or better compilers. Even more subtly, it wasn't clear whether the second algorithm was tuned to work very well on the examples that the first author had used, and perhaps it might have done worse on some other examples. So people came up with the notion of asymptotic analysis of algorithms for comparing algorithms. Also, design techniques and the notions of NP-completeness, balanced structures, efficient algorithms, and worst-case analysis were developed and added to algorithms courses.

3 External Changes Impacting Future Directions

Today, there is another fundamental change taking place. We are in a situation right now which is very similar to that of automata theory in the 1960's. The next decade or two will be every bit as exciting.

How has the computing field changed? In the 1970's , when we were thinking of programming computers, we were thinking of scientific computing. There was a belief that no matter how much computing power was available, it was never going to be enough to do simulations and so forth. But a subtle change has taken place. Most of computer usage today is not for scientific computing, and computer speed is sufficient for the kinds of things that we use computers for today: word processing, email, chat, spreadsheets, and so on. Computer usage is widespread today because you don't have to know how to program a computer in order to be able to use one. As well, there has been a merging of computing and communications: the Internet, and wireless communication. What we have to do today is rethink theoretical computer science, and develop the theory that will support these new kinds of activities.

3.1 Drivers of the Information Age

In Section 2 we covered the drivers of research in theoretical computer science back in the 1960's and 70's. Now, let's look at what is driving research today, in the so-called "information age".

Large Modifiable Systems

First, consider programming systems that are too large to be understood by any individual or small group of individuals: distributed programs consisting of many millions of lines of code, with pieces all over the world - programs that take on a life of their own, so to speak. What kind of theory can be used to support the evolution and modification of such systems by many people? Can we abstract models of cellular automata that mimic the difficulties here (there was a time in automata theory when people looked at self-organizing automata, reproducing automata, and so forth, but that work didn't lead to very much)? Or, can we abstract a graph model? Is there any concept or theory beyond directed acyclic graphs, dominators, and so forth?

Biology

Another important driver will be molecular biology. There are two interesting avenues in the biological area. One is the tremendous amount of information that we are going to acquire. The second has to do with how information is encoded in DNA, and how genes express themselves. This will drive theoretical computer science much as early work in the study of nerve sequences led to regular expressions.

Large Databases

The question here is not how to create a database (we know how to do that from work we did in the 1970's), but how do we use a database? Is there a difference between probing a database and browsing a database? Suppose, for example, you have a hard-copy of the yellow pages, and you want to look up a clothing store. You would look under "Clothing", and then you could browse around. But, if you bring up the yellow pages on the computer, you don't get the opportunity to browse around (at least on the system that I use). All you get to do is ask a question. So we might ask: is there a formal model that either proves there's a difference between probing and browsing, or proves they're really the same?

Another question concerning the use of databases is whether there is some finite set of techniques describing how we combine information together, which covers 90 percent of what we do. This is reminiscent of what we did in the design of algorithms; we looked at techniques for designing efficient algorithms which, for a wide variety of problems would give an optimal algorithm (or at least a very good one).

We should also ask whether the database information is secure. What is the concept of information? There are quite a few places now where you can find out a lot of interesting things. For example, if amazon.com keeps track of book sales by zip codes, you could ask: "what is the book most frequently bought by a Microsoft employee?" We can find out all kinds of information in subtle ways like this.

Large Graphs

Is there a model of a large graph which is analogous to finite automata? The model should be invariant under changes in the definition. That is, if there are two or three different definitions, then they all give the same set of graphs. Also, we should be able to prove some basic theorems from the definition.

For example, we need a model that will capture the web structure of the Internet, where an edge between web pages indicates that one page references the other. One of the differences between this kind of graph and the kinds of graphs that come up in graph theory, is that in graph theory, the presence of an individual edge in the graph is absolutely critical. If an edge is removed, we get a different graph, which for example may no longer have a clique of a certain size. But for the kinds of graphs that I'm interested in, which have hundreds of millions of nodes, any given edge is not particularly important. Whether or not that edge is present should not change theorems.

We need a mechanism for generating these large graphs. We need some finite mechanism which generates them, and which has certain properties which model the kinds of large graphs that we deal with.

Today, just like back in the 1960's, when all we had was E.F. Moore's selected papers, there are no textbooks describing the new theory that we need to support the the current age. What are the 8 or 9 key papers we should select to teach theory from today? If I were to teach a course in theoretical computer science within the next decade or two, I would include the paper *Authoritative sources in a hyperlinked environment* by Jon Kleinberg. He is

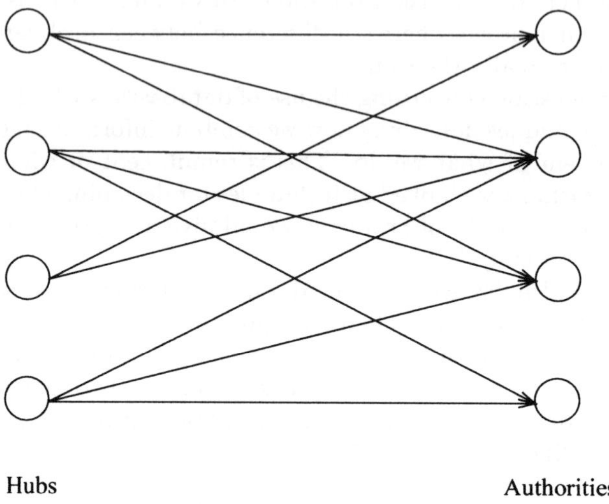

Figure 2. Kleinberg's solution

starting to make some progress in finding some definitions for large graphs. I will mention several things from his work to show why it is so important. Let's say you were going to do a search on the Internet. Say you type in the keyword *Harvard.* What is the chance you'll get Harvard's homepage? Relatively small, since Harvard's homepage doesn't have the word *Harvard* on it as frequently as many other pages do. Kleinberg gives some other examples. Suppose you want to find an automotive manufacturer. You type in *automotive manufacturer.* But you'd never get GM's homepage because *automotive manufacturer* doesn't appear on it. So how do you go about finding the pages that you want? He also points out that web pages that have a large number of pointers pointing to them aren't necessarily important ones - for example Yahoo.com has many pointers to it. So Kleinberg developed a notion of certain pages being authorities on a subject, and certain pages being hubs (see Figure 2). A link from one page to another means this page is conferring some sort of value to this other page. The idea is to extract from the web a subgraph which is bipartite, which has many hubs and many authorities, where all hubs point to the set of authorities. He observes you can actually find such subgraphs, by using the concept of an eigenvector from

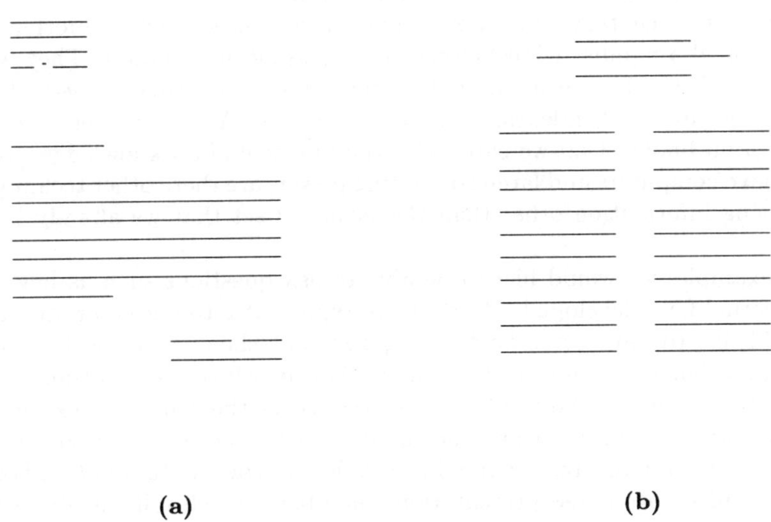

Figure 3. Document syntax

linear algebra. By doing a simple matrix calculation, you can identify a set of pages which then correspond to such a community. This might be the start of a mathematical foundation for accessing information.

Access to Information

Consider the two documents depicted in Figure 3. All of the characters have been erased from each document, but you can still probably guess that document (a) is a letter, and document (b) is an article. In other words, we can do a lot without semantics. The question is, just how much? Is syntax as good as semantics? People have discovered that syntactic search techniques can often do as well as the best semantic methods. For example, there was some work done 10 years ago by Gerald Salton on document retrieval. He used the following technique: he took a document, sorted all the words in it, and formed a vector. Similarly, he took a query, sorted all the words in it, and formed another vector. He associated with the query all documents whose vector had a dot product with the query vector that was close to one. It turned out this method of retrieval was as good as any one that tried to understand the meaning of the words in a search.

Over the past century, librarians and people who keep track of information have developed a number of techniques for accessing information. They were developed with the fact in mind that a finite amount of work is used. They have indexes, inverted indexes, and categorization. We can categorize the universe of all books. Can we categorize the universe of all knowledge? Now that we have computers and large computing power, are there other techniques for accessing information other than the standard set that we already know about?

For example, we would like to be able to ask questions such as how did an intellectual field developed. We'd like to type that into our computer, and what we'd like to have come back is a graph that shows how the key ideas developed, which ideas influenced which other ideas, who the key authors were, a map of (for example) the U.S. which shows where the centers of excellence were, how students flowed from one institution to another, and so on. In the not too distant future, we should be able to type in *Automata Theory*, and have a nice automated presentation come back which tells us about the discipline.

Intelligent Devices

We also have intelligent devices. For example, in the not too distant future, when you buy a teddy bear for a grandchild, it will probably have a 100MHz computer in it, 1 GB of memory, and be fluent in five languages. That's on the horizon. Also, we may be able to make computing technology out of plastic, or on fiber. It is conceivable that when you buy a shirt, the label in the back of the shirt will have computing power similar to the teddy bear, along with global positioning, and telecommunications (and, the label will only cost ten cents). So when I get up in the morning, look in the closet and can't find my shirt, I could just call it up (and find out for example that it's at the cleaners!). That's the world that we're moving into.

How are intelligent devices going to interact? We may have thousands of devices in our homes which are going to communicate with one another. There are over a hundred electric motors in the average house today, so we'll clearly get to the point where the average home will have over a thousand computers in it. In fact, some light switches already have computers in them, as well as infrared communication (plastic covers rather than metal). This makes rewiring older homes unnecessary, since old switches can be replaced by infrared switches.

4 Conclusions

Fundamental changes have taken place which are going to affect automata theory. We need to start building the new areas of theory and systems that are going to support these changes. It is important that people in information science and theoretical computer scientists work together at least until we find basic models. We ought to construct a book of selected works which represent the best papers available today, and start teaching new classes in theoretical computer science. Then, create a new science base for this activity, and rethink the curriculum for theoretical computer science.

References

1. Aho, A. V., J. E. Hopcroft, and J. D. Ullman [1974]. "The Design and Analysis of Computer Algorithms", Addison Wesley.
2. Hartmanis, J., and R. E. Stearns [1965]. "On the computational complexity of algorithms", *Trans. Amer. Math. Soc.* 117, 285-306.
3. Hopcroft, J. E., R. Motwani, and J. Ullman [2001]. "Introduction to Automata Theory, Languages, and Computation", 2nd ed., Addison-Wesley.
4. Kleinberg, J. [1998]. "Authoritative sources in a hyperlinked environment", Proc. 9th ACM SIAM Symposium, 668-677.
5. McNaughton, R., and H. Yamada [1960]. "Regular expressions and state graphs for automata", *IRE Trans. on Electronic Computers* 9:1, 39-47. Reprinted in Moore [1964], 157-174.
6. Moore, E. F., (ed.) [1964]. *Sequential Machines: Selected Papers*, Addison-Wesley, Reading, Mass.
7. Rabin, M. O., and D. Scott [1959]. "Finite automata and their decision problems", *IBM J. Research and Development* 3, 114-125. Reprinted in Moore [1964], 63-91.
8. Salton, Gerald [1979]. "The SMART Retrieval System - Experiments in Automatic Document Processing", Prentice Hall Inc.
9. Shepherson, J. C., [1959]. "The reduction of two-way automata to one-way automata", *IBM Journal of Research and Development*, Vol. 3, No. 2, 198-200.

FORTY YEARS OF FORMAL POWER SERIES IN AUTOMATA THEORY

WERNER KUICH
Institut für Algebra und Computermathematik
Abteilung für Theoretische Informatik
Technische Universität Wien
Wiedner Hauptstraße 8–10, A–1040 Wien
Email: kuich@tuwien.ac.at

1 Introduction

"Power from power series" (Salomaa [32]) for the theory of formal languages and automata theory was first recognized by M. P. Schützenberger at the end of the fifties. A systematic study of formal power series in noncommuting variables was initiated by Schützenberger [34]. In two seminal papers, Schützenberger [35] and Chomsky, Schützenberger [6], the theory of rational power series and algebraic power series, respectively, was developed as a generalization of the theories of regular and context-free languages, respectively. With these and other important papers, M. P. Schützenberger founded the French School on formal power series in noncommuting variables and their applications to formal languages and automata.

In the seventies A. Salomaa founded the Finnish School of formal power series in noncommutative variables culminating in the book Salomaa, Soittola [33].

Besides these two centers in the research on formal power series in noncommutative variables there emerged some singletons of researchers: S. Eilenberg in New York (in close contact with the French School), A. Paz in Haifa, W. Wechler in Dresden, S. Bozapalidis in Thessalonike, W. Kuich in Wien.

Eilenberg [7], Salomaa, Soittola [33], Paz [28], Wechler [41], Berstel, Reutenauer [3], and Kuich, Salomaa [24] are books dealing with the different aspects of formal power series in connection with formal languages and automata. Kuich [22] gives a survey on these topics, Berstel [1] contains a collection of early papers dealing with power series.

In books on automata and language theory we often have the following situation: A certain theorem is proved by a construction. In most cases the construction is intuitively clear. But the proof that the construction works is a more or less successful formalization of the construction or is left to the reader. This situation is unsatisfactory.

In this paper we hope to convince the reader that the situation becomes more satisfactory if we develop the classical results on formal languages and automata by the help of semirings, formal power series and matrices. The advantages are:

(i) The constructions needed in the proofs of the classical results are mainly the usual ones.

(ii) The descriptions of the constructions by power series and matrices are more precise than the customary ones.

(iii) The proofs are separated from the constructions and are more satisfactory from a mathematical point of view. Often they are shorter than the usual proofs.

(iv) The results are more general than the usual ones. Depending on the semiring used, the results are valid for classical automata or grammars, classical automata or grammars with ambiguity considerations, probabilistic automata or grammars, distance automata, etc.

(v) The use of formal power series and matrices gives insight into the mathematical structure of problems and yields new results and solutions to unsolved problems.

The prize to pay for these advantages is a knowledge of the basics of semiring theory.

This paper consists of this and five other sections. In Section 2, we introduce continuous semirings, formal power series and matrices, and give some rules how to perform computations in these semirings. In Section 3, automata are introduced. We first prove the Theorem of Kleene-Schützenberger. Then we show the equivalence of finite automata with and without ε-moves. In Section 4, algebraic systems and pushdown automata are introduced. They are mechanisms of equal power. The constructions and proofs given in Sections 3 and 4 have all the advantages described in items (i)–(iv) given above.

In Section 5 we give a characterization of principal cones of algebraic power series by algebraic "master systems". In the last section we mention some decidability results where no proofs with methods of language and automata theory are known. The results of Sections 5 and 6 satisfy item (v) given above.

2 Continuous monoids and semirings

In this section we first consider commutative monoids and semirings. The definitions and results on commutative monoids are mainly due—sometimes in the framework of semiring theory—to Eilenberg [7], Goldstern [11], Karner [18], Krob [19], Kuich [21,22], Kuich, Salomaa [24], Manes, Arbib [25], Sakarovitch [29]. In the second part of this section we give some identities that are valid in continuous semirings. All notions and notations that are not defined are taken from Kuich [22].

A commutative monoid $\langle A, +, 0\rangle$ is *naturally ordered* iff the set A is partially ordered by the relation $\sqsubseteq$: $a \sqsubseteq b$ iff there exists a c such that $a + c = b$.

A commutative monoid $\langle A, +, 0\rangle$ is called *complete* iff it is possible to define sums for all families $(a_i \mid i \in I)$ of elements of A, where I is an arbitrary index set, such that the following conditions are satisfied:

(i) $\sum_{i\in\emptyset} a_i = 0$, $\sum_{i\in\{j\}} a_i = a_j$, $\sum_{i\in\{j,k\}} a_i = a_j + a_k$, for $j \neq k$,

(ii) $\sum_{j\in J}(\sum_{i\in I_j} a_i) = \sum_{i\in I} a_i$, if $\bigcup_{j\in J} I_j = I$ and $I_j \cap I_{j'} = \emptyset$ for $j \neq j'$.

A complete naturally ordered monoid $\langle A, +, 0\rangle$ is called *continuous* iff for all index sets I, all families $(a_i \mid i \in I)$ in A and all $c \in A$ the following condition is satisfied:

if $\sum_{i\in E} a_i \sqsubseteq c$ for all finite subsets E of I then $\sum_{i\in I} a_i \sqsubseteq c$.

In the next theorem "sup" denotes the least upper bound with respect to the natural order.

Theorem 1 *Let $\langle A, +, 0\rangle$ be a complete naturally ordered monoid. Then A is continuous iff, for all index sets I and all families $(a_i \mid i \in I)$ in A, $\sup(\sum_{i\in E} a_i \mid E \subseteq I,\ E \text{ finite})$ exists and*

$$\sup(\sum_{i\in E} a_i \mid E \subseteq I,\ E \text{ finite}) = \sum_{i\in I} a_i.$$

Observe that $\sum_{i\in I} 0 = 0$ for an arbitrary index set I if $\langle A, +, 0\rangle$ is continuous.

An algebra $\langle A, +, \cdot, 0, 1\rangle$ is called a *semiring* iff the following conditions are satisfied for all $a, b, c \in A$:

(i) $\langle A, +, 0\rangle$ is a commutative monoid,

(ii) $\langle A, \cdot, 1\rangle$ is a monoid,

(iii) $a \cdot (b + c) = a \cdot b + a \cdot c$, $(a + b) \cdot c = a \cdot c + b \cdot c$,

(iv) $0 \cdot a = a \cdot 0 = 0$.

The semiring $\langle A, +, \cdot, 0, 1\rangle$ is called *commutative* iff $\langle A, \cdot, 1\rangle$ is a commutative monoid. It is called *naturally ordered* iff $\langle A, +, 0\rangle$ is naturally ordered. It is called *complete* iff $\langle A, +, 0\rangle$ is complete and the following additional conditions are satisfied for all index sets I and all $a, a_i \in A$, $i \in I$:

$$a \cdot (\sum_{i\in I} a_i) = \sum_{i\in I} a \cdot a_i, \qquad (\sum_{i\in I} a_i) \cdot a = \sum_{i\in I} a_i \cdot a.$$

A complete semiring $\langle A, +, \cdot, 0, 1\rangle$ is called *continuous* iff $\langle A, +, 0\rangle$ is continuous.

Intuitively, a semiring is a ring (with unity) without subtraction. A typical example is the semiring of nonnegative integers $\mathbb{N}$. A very important semiring in connection with language theory is the *Boolean* semiring $\mathbb{B} = \{0, 1\}$ where $1 + 1 = 1 \cdot 1 = 1$. Clearly, all rings (with unity), as well as all fields, are semirings, e. g., integers $\mathbb{Z}$, rationals $\mathbb{Q}$, reals $\mathbb{R}$, complex numbers $\mathbb{C}$ etc. Let $\mathbb{N}^\infty = \mathbb{N} \cup \{\infty\}$. Then $\langle \mathbb{N}^\infty, +, \cdot, 0, 1\rangle$ and $\langle \mathbb{N}^\infty, \min, +, \infty, 0\rangle$, where $+$, $\cdot$ and min are defined in the obvious fashion (observe that $0 \cdot \infty = \infty \cdot 0 = 0$), are semirings. The first one is usually denoted by $\mathbb{N}^\infty$, the second one is called the *tropical* semiring. Let $\mathbb{R}_+ = \{a \in \mathbb{R} \mid a \geq 0\}$ and $\mathbb{R}_+^\infty = \mathbb{R}_+ \cup \{\infty\}$. Then $\langle \mathbb{R}_+, +, \cdot, 0, 1\rangle$ and $\langle \mathbb{R}_+^\infty, +, \cdot, 0, 1\rangle$ are semirings. The semirings $\mathbb{B}$, $\langle \mathbb{N}^\infty, +, \cdot, 0, 1\rangle$, $\langle \mathbb{N}^\infty, \min, +, \infty, 0\rangle$ and $\mathbb{R}_+^\infty$ are continuous semirings.

Let Σ be an alphabet and define, for $L_1, L_2 \subseteq \Sigma^*$, the *product* of L_1 and L_2 by

$$L_1 \cdot L_2 = \{w_1 w_2 \mid w_1 \in L_1,\ w_2 \in L_2\}.$$

Then $\langle \mathfrak{P}(\Sigma^*), \cup, \cdot, \emptyset, \{\varepsilon\}\rangle$ is a continuous semiring. In case Σ is finite it is called the *semiring of formal languages over* Σ. Here $\mathfrak{P}(S)$ denotes the power set of a set S and $\emptyset$ denotes the empty set.

Given a continuous semiring A, an alphabet Σ and a nonempty index set I, we will introduce two more continuous semirings: The semiring $A\langle\langle\Sigma^*\rangle\rangle$ of formal power series (over Σ^*) and the semiring $A^{I\times I}$ of matrices (indexed by $I \times I$).

We start with formal power series. Mappings r from Σ^* into A are called *(formal) power series*. The values of r are denoted by (r, w), where $w \in \Sigma^*$, and r itself is written as a formal sum

$$r = \sum_{w\in\Sigma^*} (r, w)w.$$

The values (r, w) are also referred to as the *coefficients* of the series. The collection of all power series r as defined above is denoted by $A\langle\langle\Sigma^*\rangle\rangle$.

Given $r \in A\langle\langle\Sigma^*\rangle\rangle$, the subset of Σ^* defined by

$$\{w \mid (r, w) \neq 0\}$$

is termed the *support* of r and denoted by $\text{supp}(r)$. The subset of $A\langle\langle\Sigma^*\rangle\rangle$ consisting of all series with a finite support is denoted by $A\langle\Sigma^*\rangle$. Series of $A\langle\Sigma^*\rangle$ are referred to as *polynomials.*

Examples of polynomials belonging to $A\langle\Sigma^*\rangle$ for every A are 0, w, aw, $a \in A$, $w \in \Sigma^*$, defined by:

$$\begin{aligned}
&(0, w) = 0 \text{ for all } w,\\
&(w, w) = 1 \text{ and } (w, w') = 0 \text{ for } w \neq w',\\
&(aw, w) = a \text{ and } (aw, w') = 0 \text{ for } w \neq w'.
\end{aligned}$$

We now introduce two operations inducing a semiring structure to power series. For $r_1, r_2 \in A\langle\langle\Sigma^*\rangle\rangle$, we define the *sum* $r_1 + r_2 \in A\langle\langle\Sigma^*\rangle\rangle$ by $(r_1 + r_2, w) = (r_1, w) + (r_2, w)$ for all $w \in \Sigma^*$. For $r_1, r_2 \in A\langle\langle\Sigma^*\rangle\rangle$, we define the *(Cauchy) product* $r_1 r_2 \in A\langle\langle\Sigma^*\rangle\rangle$ by $(r_1 r_2, w) = \sum_{w_1 w_2 = w}(r_1, w_1)(r_2, w_2)$ for all $w \in \Sigma^*$. Clearly, $\langle A\langle\langle\Sigma^*\rangle\rangle, +, \cdot, 0, \varepsilon\rangle$ and $\langle A\langle\Sigma^*\rangle, +, \cdot, 0, \varepsilon\rangle$ are semirings.

Moreover, $A\langle\langle\Sigma^*\rangle\rangle$ is a continuous semiring. Sums in $A\langle\langle\Sigma^*\rangle\rangle$ are defined by $(\sum_{j \in J} r_j, w) = \sum_{j \in J}(r_j, w)$ for all $w \in \Sigma^*$ and all index sets J.

For $a \in A$, $r \in A\langle\langle\Sigma^*\rangle\rangle$, we define the *scalar products* $ar, ra \in A\langle\langle\Sigma^*\rangle\rangle$ by $(ar, w) = a(r, w)$ and $(ra, w) = (r, w)a$ for all $w \in \Sigma^*$. Observe that $ar = (a\varepsilon)r$ and $ra = r(a\varepsilon)$. If A is commutative then $ar = ra$.

A series $r \in A\langle\langle\Sigma^*\rangle\rangle$, where every coefficient equals 0 or 1, is termed the *characteristic series* of its support L, in symbols, $r = \text{char}(L)$.

It will be convenient to use the notations $A\langle\Sigma \cup \varepsilon\rangle$, $A\langle\Sigma\rangle$ and $A\langle\varepsilon\rangle$ for the collection of polynomials having their supports in $\Sigma \cup \{\varepsilon\}$, Σ and $\{\varepsilon\}$, respectively.

Clearly, $\mathfrak{P}(\Sigma^*)$ is a semiring isomorphic to $\mathbb{B}\langle\langle\Sigma^*\rangle\rangle$. Essentially, a transition from $\mathfrak{P}(\Sigma^*)$ to $\mathbb{B}\langle\langle\Sigma^*\rangle\rangle$ and vice versa means a transition from L to $\text{char}(L)$ and from r to $\text{supp}(r)$, respectively.

We now introduce matrices. Consider two nonempty index sets I and I'. Mappings M of $I \times I'$ into A are called *matrices.* The values of M are denoted by $M_{i,i'}$, where $i \in I$ and $i' \in I'$. The values $M_{i,i'}$ are also referred to as the *entries* of the matrix M. In particular, $M_{i,i'}$ is called the (i, i')*-entry* of M. The collection of all matrices as defined above is denoted by $A^{I \times I'}$. Let $A' \subseteq A$. Then the collection of all matrices in $A^{I \times I'}$ that have their entries in A' is denoted by $A'^{I \times I'}$.

If I or I' is a singleton, M is called a *row* or *column vector*, respectively. If both I and I' are finite, then M is called a *finite matrix.*

We introduce some operations and special matrices inducing a monoid or semiring structure to matrices. For $M_1, M_2 \in A^{I\times I'}$ we define the *sum* $M_1 + M_2 \in A^{I\times I'}$ by $(M_1 + M_2)_{i,i'} = (M_1)_{i,i'} + (M_2)_{i,i'}$ for all $i \in I$, $i' \in I'$. Furthermore, we introduce the *zero matrix* $0 \in A^{I\times I'}$. All entries of the zero matrix 0 are 0. By these definitions, $\langle A^{I\times I'}, +, 0\rangle$ is a commutative monoid.

For $M_1 \in A^{I_1\times I_2}$ and $M_2 \in A^{I_2\times I_3}$ we define the *product* $M_1M_2 \in A^{I_1\times I_3}$ by

$$(M_1M_2)_{i_1,i_3} = \sum_{i_2\in I_2} (M_1)_{i_1,i_2}(M_2)_{i_2,i_3} \qquad \text{for all } i_1 \in I_1,\ i_3 \in I_3.$$

Furthermore, we introduce the *matrix of unity* $E \in A^{I\times I}$. The diagonal entries $E_{i,i}$ of E are equal to 1, the off-diagonal entries E_{i_1,i_2}, $i_1 \neq i_2$, of E are equal to 0, $i, i_1, i_2 \in I$.

It is easily shown that matrix multiplication is associative, the distribution laws are valid for matrix addition and multiplication, E is a multiplicative unit and 0 is a multiplicative zero. So we infer that $\langle A^{I\times I}, +, \cdot, 0, E\rangle$ is a semiring.

Infinite sums can be extended to matrices. Consider $A^{I\times I'}$ and define, for $M_j \in A^{I\times I'}$, $j \in J$, where J is an index set, $\sum_{j\in J} M_j$ by its entries:

$$\Big(\sum_{j\in J} M_j\Big)_{i,i'} = \sum_{j\in J} (M_j)_{i,i'}, \qquad i \in I,\ i' \in I'.$$

By this definition, $A^{I\times I}$ is an continuous semiring.

Moreover, the natural order on A is extended entrywise to matrices M_1 and M_2 in $A^{I\times I'}$:

$$M_1 \sqsubseteq M_2 \qquad \text{iff} \qquad (M_1)_{i,i'} \sqsubseteq (M_2)_{i,i'} \text{ for all } i \in I,\ i' \in I'.$$

Here and in the whole paper, A denotes a *continuous* semiring, I denotes a nonempty index set and Σ a finite alphabet. Moreover, Q will denote a finite nonempty index set. All notational letters may be indexed.

The *star* a^* of $a \in A$ is defined by $a^* = \sum_{n\geq 0} a^n$. We need two results that give us rules how to perform computations.

Theorem 2 $(a + b)^* = a^*(ba^*)^* = (a^*b)^*a^* = (a + ba^*b)^*(1 + ba^*)$ *for all* $a, b \in A$.

In the next corollary, $M^*(I_i, I_j)$ and $M_{i,j}$, $1 \leq i, j \leq 2$, denote the blocks of the matrices M^* and M, respectively.

Corollary 3 *Let* $M \in A^{I\times I}$ *and* $I = I_1 \cup I_2$, $I_1 \cap I_2 = \emptyset$. *Then*

$$M^*(I_1, I_1) = (M_{1,1} + M_{1,2}M_{2,2}^*M_{2,1})^*,$$
$$M^*(I_1, I_2) = (M_{1,1} + M_{1,2}M_{2,2}^*M_{2,1})^*M_{1,2}M_{2,2}^*,$$

$$M^*(I_2, I_1) = (M_{2,2} + M_{2,1}M_{1,1}^*M_{1,2})^* M_{2,1}M_{1,1}^*,$$
$$M^*(I_2, I_2) = (M_{2,2} + M_{2,1}M_{1,1}^*M_{1,2})^*.$$

Proof. Let $M', M'' \in A^{I \times I}$ be defined by their blocks:

$$M'_{1,1} = M_{1,1},\ M'_{1,2} = 0, \qquad M'_{2,1} = 0, \qquad M'_{2,2} = M_{2,2};$$
$$M''_{1,1} = 0, \qquad M''_{1,2} = M_{1,2},\ M''_{2,1} = M_{2,1},\ M''_{2,2} = 0.$$

Then $M = M' + M''$. Apply now Theorem 2.

We call $+$, $\cdot$, and * the *rational operations.* A subsemiring of A is called *rationally closed* iff it is closed under rational operations. By definition, $\mathfrak{Rat}(A')$ is the smallest rationally closed subsemiring of A containing the subset A' of A.

Corollary 3 shows that the blocks of M^* are computed by applying rational operations on the blocks of M. Hence, a simple proof by induction yields the next theorem.

Theorem 4 *Let A' be a rationally closed subsemiring of A. Then*

$$M \in A'^{Q \times Q} \qquad \textit{implies} \qquad M^* \in A'^{Q \times Q}.$$

3 Automata and the Theorem of Kleene

The well-known finite automata are generalized in the following two directions:

(i) An infinite set of states will be allowed in the general definition. When dealing with pushdown automata in Section 4 this will enable us to store the contents of the pushdown tape in the states.

(ii) A single state transition generates an element of $A\langle \Sigma \cup \varepsilon \rangle$ instead of reading a symbol. Thus an automaton generates an element of the basic semiring $A\langle\langle \Sigma^* \rangle\rangle$.

Our model of an automaton will be defined in terms of a (possibly infinite) transition matrix. The semiring element generated by the transition of the automaton from one state i to another state i' in exactly k computation steps equals the (i, i')-entry in the k-th power of the transition matrix. Consider now the star of the transition matrix. Then the semiring element generated by the automaton, also called the behavior of the automaton, can be expressed by the entries (multiplied by the initial and final weights of the states) of the star of the transition matrix.

An *automaton* (*with input alphabet Σ over the semiring A*)

$$\mathfrak{A} = (I, M, S, P)$$

is given by

(i) a nonempty set I of *states*,

(ii) a matrix $M \in (A\langle \Sigma \cup \varepsilon \rangle)^{I \times I}$, called the *transition matrix*,

(iii) $S \in A\langle \varepsilon \rangle^{1 \times I}$, called the *initial state vector*,

(iv) $P \in A\langle \varepsilon \rangle^{I \times 1}$, called the *final state vector*.

The *behavior* $\|\mathfrak{A}\| \in A$ of the automaton $\mathfrak{A}$ is defined by

$$\|\mathfrak{A}\| = \sum_{i_1, i_2 \in I} S_{i_1} (M^*)_{i_1, i_2} P_{i_2} = SM^*P.$$

An automaton is termed *finite* iff its state set is finite. We define

$$\mathfrak{E}(\Sigma) = \{\|\mathfrak{A}\| \mid \mathfrak{A} \text{ is a finite automaton with input alphabet } \Sigma\},$$

and $A^{\mathrm{rat}}\langle\langle \Sigma^* \rangle\rangle = \mathfrak{Rat}(A\langle \Sigma^* \rangle)$. The power series in $A^{\mathrm{rat}}\langle\langle \Sigma^* \rangle\rangle$ are called *rational power series*. Theorem 4 implies $\mathfrak{E}(\Sigma) \subseteq A^{\mathrm{rat}}\langle\langle \Sigma^* \rangle\rangle$. We will show the converse containment in Theorem 6.

By definition, an automaton $\mathfrak{A} = (I, M, S, P)$ is *normalized* iff the following conditions (i), (ii) and (iii) are satisfied.

(i) There exists an $i_0 \in I$ such that $S_{i_0} = \varepsilon$ and $S_i = 0$ for $i \neq i_0$.

(ii) There exists an $i_f \in I$, $i_f \neq i_0$, such that $P_{i_f} = \varepsilon$ and $P_i = 0$ for $i \neq i_f$.

(iii) $M_{i,i_0} = M_{i_f,i} = 0$ for all $i \in I$.

Theorem 5 *If r is the behavior of an automaton then r is also the behavior of a normalized automaton.*

Proof. Assume $r = \|\mathfrak{A}\|$, where $\mathfrak{A} = (I, M, S, P)$ is an automaton. Let i_0 and i_f be new states and define the automaton $\mathfrak{A}' = (I \cup \{i_0, i_f\}, M', S', P')$, where

$$S' = \begin{pmatrix} \varepsilon & 0 & 0 \end{pmatrix}, \quad M' = \begin{pmatrix} 0 & S & 0 \\ 0 & M & P \\ 0 & 0 & 0 \end{pmatrix}, \quad P' = \begin{pmatrix} 0 \\ 0 \\ \varepsilon \end{pmatrix}.$$

(Here the blocks are indexed by i_0, I and i_f.) By Corollary 3, we obtain $(M'^*)_{i_0,i_f} = SM^*P$. Hence, $\|\mathfrak{A}'\| = S'M'^*P' = (M'^*)_{i_0,i_f} = SM^*P = \|\mathfrak{A}\|$.

Theorem 6 (Theorem of Kleene-Schützenberger [35])

$$\mathfrak{E}(\Sigma) = A^{\mathrm{rat}}\langle\langle \Sigma^* \rangle\rangle.$$

Proof. We have to show that $\mathfrak{C}(\Sigma)$ is a rationally closed semiring containing $A\langle\Sigma^*\rangle$, i. e., we have to show closure under $+$, $\cdot$ and *.

Let $\mathfrak{A} = (Q, M, S, P)$ and $\mathfrak{A}' = (Q', M', S', P')$ be finite automata, where Q and Q' are disjoint. Then we construct finite automata $\mathfrak{A}_1$, $\mathfrak{A}_2$ and $\mathfrak{A}_3$ with behaviors $\|\mathfrak{A}\| + \|\mathfrak{A}'\|$, $\|\mathfrak{A}\| \cdot \|\mathfrak{A}'\|$ and $\|\mathfrak{A}\|^*$, respectively.

(i) Define $\mathfrak{A}_1 = (Q \cup Q', M_1, S_1, P_1)$ by

$$M_1 = \begin{pmatrix} M & 0 \\ 0 & M' \end{pmatrix}, \quad S_1 = \begin{pmatrix} S & S' \end{pmatrix}, \quad P_1 = \begin{pmatrix} P \\ P' \end{pmatrix}.$$

(ii) Define $\mathfrak{A}_2 = (Q \cup Q', M_2, S_2, P_2)$ by

$$M_2 = \begin{pmatrix} M & PS' \\ 0 & M' \end{pmatrix}, \quad S_2 = \begin{pmatrix} S & 0 \end{pmatrix}, \quad P_2 = \begin{pmatrix} 0 \\ P' \end{pmatrix}.$$

(iii) Define $\mathfrak{A}_3 = (Q \cup \{q_0\}, M_3, S_3, P_3)$, where q_0 is a new state, by

$$M_3 = \begin{pmatrix} 0 & S \\ P & M \end{pmatrix}, \quad S_3 = \begin{pmatrix} \varepsilon & 0 \end{pmatrix}, \quad P_3 = \begin{pmatrix} \varepsilon \\ 0 \end{pmatrix}.$$

Corollary 3 yields at once $\|\mathfrak{A}_1\| = \|\mathfrak{A}\| + \|\mathfrak{A}'\|$, $\|\mathfrak{A}_2\| = \|\mathfrak{A}\| \cdot \|\mathfrak{A}'\|$ and $\|\mathfrak{A}_3\| = (SM^*P)^* = \|\mathfrak{A}\|^*$.

For each ax, $a \in A$, $x \in \Sigma$ a trivial construction yields a finite automaton whose behavior is ax. Hence, constructions (i) and (ii) show that the set of behaviors of finite automata forms a semiring containing $A\langle\Sigma^*\rangle$. Construction (iii) shows that this semiring is rationally closed.

We now show a result that allows us to delete ε-transitions without changing the behavior of an automaton. (See also Hopcroft, Ullman [16], Theorem 2.2.)

Theorem 7 *Consider an automaton $\mathfrak{A} = (I, M, S, P)$. Then there exists an automaton $\mathfrak{A}' = (I, M', S, P')$, where $M' \in (A\langle\Sigma\rangle)^{I\times I}$, such that $\|\mathfrak{A}'\| = \|\mathfrak{A}\|$.*

Proof. Define $M_0 = (M, \varepsilon)\varepsilon$, $M_1 = \sum_{x\in\Sigma}(M, x)x$ and let $M' = M_0^* M_1$, $P' = M_0^* P$. Then, by Theorem 2,

$$\|\mathfrak{A}'\| = S(M_0^* M_1)^* M_0^* P = S(M_0 + M_1)^* P = SM^*P = \|\mathfrak{A}\|.$$

In the next corollary we use the isomorphism between $\mathfrak{P}(\Sigma^*)$ and $\mathbb{B}\langle\langle\Sigma^*\rangle\rangle$.

Corollary 8 (Theorem of Kleene [20]) *The following statements on a formal language L over Σ are equivalent:*

(i) *L is the behavior of a finite automaton $\mathfrak{A} = (Q, M, S, P)$ over the semiring $\mathbb{B}$ such that $M \in \mathbb{B}\langle\Sigma\rangle^{Q\times Q}$ and $S_{q_0} = \varepsilon$, $S_q = 0$, $q \neq q_0$, for some $q_0 \in Q$.*

(ii) $L \in \mathbb{B}^{\mathrm{rat}}\langle\langle\Sigma^*\rangle\rangle$.

In language theory, a formula telling how a given regular language is obtained from the languages $\{x\}$, $x \in \Sigma$, and $\emptyset$ by rational operations is referred to as a *regular expression.* Hence, item (ii) of Corollary 8 means that L is denoted by a regular expression (see Salomaa [30] and [31], Theorem 5.1).

4 Algebraic systems and pushdown automata

In the rest of this paper, A denotes a *commutative* continuous semiring and $Y = \{y_1, \ldots, y_n\}$ a set of variables, $Y \cap \Sigma = \emptyset$. An *algebraic system (with variables in* Y *over* $A\langle\langle\Sigma^*\rangle\rangle$) is a system of equations

$$y_i = p_i, \quad 1 \leq i \leq n,$$

where each p_i is a polynomial in $A\langle(\Sigma \cup Y)^*\rangle$.

Intuitively, a solution of the algebraic system $y_i = p_i(y_1, \ldots, y_n)$, $1 \leq i \leq n$, is given by $(\sigma_1, \ldots, \sigma_n) \in (A\langle\langle\Sigma^*\rangle\rangle)^n$ "satisfying" the algebraic system in the sense that if each variable y_i is replaced by the series σ_i then valid equations result, i. e., $\sigma_i = p_i(\sigma_1, \ldots, \sigma_n)$, $1 \leq i \leq n$.

More formally, consider $\sigma = (\sigma_1, \ldots, \sigma_n) \in (A\langle\langle\Sigma^*\rangle\rangle)^n$. Then we can define a morphism $\bar{\sigma} : (\Sigma \cup Y)^* \to A\langle\langle\Sigma^*\rangle\rangle$ by $\bar{\sigma}(y_i) = \sigma_i$, $1 \leq i \leq n$, and $\bar{\sigma}(x) = x$, $x \in \Sigma$, and extend it to a semiring morphism $\bar{\sigma} : A\langle\langle(\Sigma \cup Y)^*\rangle\rangle \to A\langle\langle\Sigma^*\rangle\rangle$ by the definition

$$\bar{\sigma}(r) = \sum_{\alpha \in (\Sigma \cup Y)^*} (r, \alpha)\bar{\sigma}(\alpha), \quad r \in A\langle\langle(\Sigma \cup Y)^*\rangle\rangle .$$

We denote now $\bar{\sigma}(r(y_1, \ldots, y_n))$ by $r(\sigma_1, \ldots, \sigma_n)$ and say that $r(\sigma_1, \ldots, \sigma_n)$ originates from $r(y_1, \ldots, y_n)$ by *substituting* σ_i for y_i, $1 \leq i \leq n$.

A *solution* to the algebraic system $y_i = p_i(y_1, \ldots, y_n)$, $1 \leq i \leq n$, is now given by $(\sigma_1, \ldots, \sigma_n) \in (A\langle\langle\Sigma^*\rangle\rangle)^n$ such that $\sigma_i = p_i(\sigma_1, \ldots, \sigma_n)$, $1 \leq i \leq n$. A solution $(\sigma_1, \ldots, \sigma_n)$ of the algebraic system $y_i = p_i$, $1 \leq i \leq n$, is termed *least solution* iff $\sigma_i \sqsubseteq \tau_i$, $1 \leq i \leq n$, for all solutions $(\tau_1, \ldots, \tau_n)$ of $y_i = p_i$, $1 \leq i \leq n$. Since the least solution of an algebraic system $y_i = p_i$, $1 \leq i \leq n$ is nothing else than the least fixpoint of the mapping induced by $(p_1(y_1, \ldots, y_n), \ldots, p_n(y_1, \ldots, y_n))$ and since this mapping is continuous we can apply the Fixpoint Theorem (see Wechler [41], Kuich [22]).

Theorem 9 *The least solution of an algebraic system exists in* $(A\langle\langle\Sigma^*\rangle\rangle)^n$.

The collection of the components of the least solutions of all algebraic systems over $A\langle\langle\Sigma^*\rangle\rangle$ is denoted by $A^{\mathrm{alg}}\langle\langle\Sigma^*\rangle\rangle$. The power series in $A^{\mathrm{alg}}\langle\langle\Sigma^*\rangle\rangle$ are called *algebraic power series.*

We now discuss the connection between algebraic systems and context-free grammars. Define, for a given context-free grammar $G = (Y, \Sigma, P, y_1)$, the algebraic system $y_i = p_i$, $p_i \in A\langle(\Sigma \cup Y)^*\rangle$, $1 \le i \le n$, by

$$(p_i, \gamma) = 1 \text{ if } y_i \to \gamma \in P \quad \text{and} \quad (p_i, \gamma) = 0, \text{ otherwise.}$$

Conversely, given an algebraic system $y_i = p_i$, $p_i \in A\langle(\Sigma \cup Y)^*\rangle$, $1 \le i \le n$, define the context-free grammar $G = (Y, \Sigma, P, y_1)$ by

$$y_i \to \gamma \in P \quad \text{iff} \quad (p_i, \gamma) \ne 0.$$

Whenever we speak of a context-free grammar corresponding to an algebraic system $y_i = p_i$, $p_i \in A\langle(\Sigma \cup Y)^*\rangle$, $1 \le i \le n$, or vice versa, then we mean the correspondence in the sense of the above definition. If attention is restricted to algebraic systems with coefficients 0 and 1 then this correspondence is one-to-one. Hence, the correspondence between algebraic systems over $\mathbb{B}\langle\langle\Sigma^*\rangle\rangle$ and context-free grammars is one-to-one.

The next theorem is due to Ginsburg, Rice [10]. (See also Salomaa, Soittola [33], Theorem IV.1.2 and Moll, Arbib, Kfoury [26], Chapter 6.)

Theorem 10 (Ginsburg, Rice [10], Theorem 2) *Assume that $G = (Y, \Sigma, P, y_1)$ is a context-free grammar and $y_i = p_i$, $1 \le i \le n$, is the corresponding algebraic system over $\mathbb{B}\langle\langle\Sigma^*\rangle\rangle$ with least solution $(\sigma_1, \ldots, \sigma_n)$. Let $G_i = (Y, \Sigma, P, y_i)$, $1 \le i \le n$. Then*

$$L(G_i) = \operatorname{supp}(\sigma_i), \quad 1 \le i \le n.$$

Corollary 11 *A formal language is context-free iff it is the support of an algebraic power series.*

The next theorem is due to Chomsky, Schützenberger [6].

Theorem 12 (Salomaa, Soittola [33], Theorem IV.1.5) *Assume that $G = (Y, \Sigma, P, y_1)$ is a context-free grammar and $y_i = p_i$, $p_i \in \mathbb{N}^\infty\langle(\Sigma \cup Y)^*\rangle$, $1 \le i \le n$, is the corresponding algebraic system with least solution $(\sigma_1, \ldots, \sigma_n)$. Denote by $d_i(w)$ the number (possibly ∞) of distinct leftmost derivations of w from the variable y_i, $1 \le i \le n$, $w \in \Sigma^*$. Then*

$$\sigma_i = \sum_{w \in \Sigma^*} d_i(w) w, \quad 1 \le i \le n.$$

Corollary 13 *Under the assumptions of Theorem 12, G is unambiguous iff, for all $w \in \Sigma^*$,*

$$(\sigma_1, w) \le 1 .$$

We now define pushdown automata. Pushdown automata are finite automata (with state set Q) augmented by a pushdown tape. The contents of the pushdown tape is a word over the pushdown alphabet Γ. We consider a pushdown automaton to be an automaton in the sense of Section 3: the state set is given by $\Gamma^* \times Q$ and its transition matrix is in $A\langle\Sigma \cup \varepsilon\rangle^{(\Gamma^* \times Q) \times (\Gamma^* \times Q)}$. This allows us to store the contents of the pushdown tape and the states of the finite automaton in the states of the pushdown automaton. Because of technical reasons, we do not work in the semiring $A\langle\langle\Sigma^*\rangle\rangle^{(\Gamma^* \times Q) \times (\Gamma^* \times Q)}$ but in the isomorphic semiring $(A\langle\langle\Sigma^*\rangle\rangle^{Q \times Q})^{\Gamma^* \times \Gamma^*}$. A matrix $M \in (A\langle\Sigma \cup \varepsilon\rangle^{Q \times Q})^{\Gamma^* \times \Gamma^*}$ is termed a *pushdown transition matrix* iff

(i) for each $p \in \Gamma$ there exist only finitely many blocks $M_{p,\pi}$, $\pi \in \Gamma^*$, that are unequal to 0;

(ii) for all $\pi_1, \pi_2 \in \Gamma^*$,

$$M_{\pi_1,\pi_2} = \begin{cases} M_{p,\pi} & \text{if there exist } p \in \Gamma,\ \pi' \in \Gamma^* \text{ with} \\ & \pi_1 = p\pi' \text{ and } \pi_2 = \pi\pi', \\ 0 & \text{otherwise.} \end{cases}$$

The above definition implies that a pushdown transition matrix has a finitary specification: it is completely specified by its non-null blocks of the form $M_{p,\pi}$, $p \in \Gamma$, $\pi \in \Gamma^*$. Item (ii) of the above definition shows that only the following transitions are possible: if the contents of the pushdown tape is given by $p\pi'$, the contents of the pushdown tape after a transition has to be of the form $\pi\pi'$; moreover, the transition does only depend on the leftmost (topmost) pushdown sympol p and not on π'. In this sense the pushdown transition matrix represents a proper formalization of the principle "last in—first out".

A *pushdown automaton* (*with input alphabet* Σ *over the semiring* A)

$$\mathfrak{P} = (Q, \Gamma, M, S, p_0, P)$$

is given by

(i) a finite set Q of *states*,

(ii) a finite alphabet Γ of *pushdown symbols*,

(iii) a *pushdown transition matrix* $M \in (A\langle\Sigma \cup \varepsilon\rangle^{Q \times Q})^{\Gamma^* \times \Gamma^*}$,

(iv) $S \in A\langle\varepsilon\rangle^{1 \times Q}$, called the *initial state vector*,

(v) $p_0 \in \Gamma$, called the *initial pushdown symbol*,

(vi) $P \in A\langle\varepsilon\rangle^{Q \times 1}$, called the *final state vector*.

The behavior $\|\mathfrak{P}\|$ of the pushdown automaton $\mathfrak{P}$ is defined by

$$\|\mathfrak{P}\| = S(M^*)_{p_0,\varepsilon}P.$$

We now describe the computations of a pushdown automaton. Initially, the pushdown tape contains the special symbol p_0. The pushdown automaton now performs transitions governed by the pushdown transition matrix until the pushdown tape is emptied. The result of these computations is given by $(M^*)_{p_0,\varepsilon}$. Multiplications by the initial state vector and by the final state vector yield the behavior of the pushdown automaton.

Let now $\mathbb{B}\langle\langle\Sigma^*\rangle\rangle$ be our basic semiring. We connect our definition of a *pushdown automaton* $\mathfrak{P} = (Q, \Gamma, M, S, p_0, P)$ *over* $\mathbb{B}$ to the usual definition of a *pushdown automaton* $\mathfrak{P}' = (Q, \Sigma, \Gamma, \delta, q_0, p_0, F)$ (see e. g., Harrison [14]), where Σ is the *input alphabet*, $\delta : Q \times (\Sigma \cup \{\varepsilon\}) \times \Gamma \to$ finite subsets of $Q \times \Gamma^*$ is the *transition function*, $q_0 \in Q$ is the *initial state* and $F \subseteq Q$ is the set of *final states.*

Assume that a pushdown automaton $\mathfrak{P}'$ is given as above. The transition function δ defines the pushdown transition matrix M of $\mathfrak{P}$ by

$$((M_{p,\pi})_{q_1,q_2}, x) = 1 \quad \text{iff} \quad (q_2, \pi) \in \delta(q_1, x, p)$$

for all $q_1, q_2 \in Q$, $p \in \Gamma$, $\pi \in \Gamma^*$, $x \in \Sigma \cup \{\varepsilon\}$. Let now $\vdash$ be the move relation over the *instantaneous descriptions* of $\mathfrak{P}'$ in $Q \times \Sigma^* \times \Gamma^*$. Then $(q_1, w, \pi_1) \vdash^k (q_2, \varepsilon, \pi_2)$ iff $(((M^k)_{\pi_1,\pi_2})_{q_1,q_2}, w) = 1$ and $(q_1, w, \pi_1) \vdash^* (q_2, \varepsilon, \pi_2)$ iff $(((M^*)_{\pi_1,\pi_2})_{q_1,q_2}, w) = 1$ for all $k \geq 0$, $q_1, q_2 \in Q$, $\pi_1, \pi_2 \in \Gamma^*$, $w \in \Sigma^*$. Hence, $(q_0, w, p_0) \vdash^* (q, \varepsilon, \varepsilon)$ iff $(((M^*)_{p_0,\varepsilon})_{q_0,q}, w) = 1$. Define the initial state vector S and the final state vector P by $S_{q_0} = \varepsilon$, $S_q = 0$ if $q \neq q_0$, $P_q = \varepsilon$ if $q \in F$, $P_q = 0$ if $q \notin F$. Then a word w is accepted by the pushdown automaton $\mathfrak{P}'$ by both final state and empty store iff $(S(M^*)_{p_0,\varepsilon}P, w) = (\|\mathfrak{P}\|, w) = 1$.

Let $\mathfrak{P} = (Q, \Gamma, M, S, p_0, P)$ be a pushdown automaton and consider the algebraic system

$$\begin{aligned} y_0 &= S y_{p_0} P, \\ y_p &= \sum_{\pi \in \Gamma^*} M_{p,\pi} y_\pi, \quad p \in \Gamma, \end{aligned}$$

written in matrix notation: y_p is a $Q \times Q$-matrix whose (q_1, q_2)-entry is the variable $[q_1, p, q_2]$, $p \in \Gamma$, $q_1, q_2 \in Q$; if $\pi = p_1 \ldots p_r$, $r \geq 1$, then the (q_1, q_2)-entry of y_π is given by the (q_1, q_2)-entry of $y_{p_1} \ldots y_{p_r}$, $p_1, \ldots, p_r \in \Gamma$; if $\pi = \varepsilon$ then $y_\pi = E$; y_0 is a variable. Hence, the variables of the above algebraic system are y_0, $[q_1, p, q_2]$, $p \in \Gamma$, $q_1, q_2 \in Q$.

Theorem 14 (Kuich [22], Corollary 6.5) *Let* $\mathfrak{P} = (Q, \Gamma, M, S, p_0, P)$ *be a pushdown automaton. Then* $\|\mathfrak{P}\|$, $(M^*)_{p,\varepsilon}$, $p \in \Gamma$, *is the least solution of the algebraic system*

$$y_0 = S y_{p_0} P,$$
$$y_p = \sum_{\pi \in \Gamma^*} M_{p,\pi} y_\pi, \quad p \in \Gamma.$$

Observe that in case of the Boolean semiring $\mathbb{B}$ the construction of Theorem 14 is nothing else than the well-known triple construction. (See Hopcroft, Ullman [16], Theorem 5.4; Harrison [14], Theorem 5.4.3; Bucher, Maurer [4], Satz 2.3.10.) It is clear that $[q_1, p, q_2] \Rightarrow^* w$ in G iff $(((M^*)_{p,\varepsilon})_{q_1,q_2}, w) = 1$ in $\mathfrak{P}$ iff $(q_1, w, p) \vdash^* (q_2, \varepsilon, \varepsilon)$ in $\mathfrak{P}'$, $q_1, q_2 \in Q$, $p \in \Gamma$, $w \in \Sigma^*$. This means that there exists a derivation of w from the variable $[q_1, p, q_2]$ iff w empties the pushdown tape with contents p by a computation from state q_1 to state q_2. The interested reader should compare the proof of Corollary 6.5 in Kuich [22] with the usual proofs of the triple construction.

If our basic semiring is $\mathbb{N}^\infty \langle\langle \Sigma^* \rangle\rangle$, we can draw some even stronger conclusions by Theorem 12. In Theorem 15 we consider, for a given pushdown automaton $\mathfrak{P}' = (Q, \Sigma, \Gamma, \delta, q_0, p_0, F)$, the number of distinct computations from the *initial instantaneous description* (q_0, w, p_0) *for* w to an *accepting instantaneous description* $(q, \varepsilon, \varepsilon)$, $q \in F$.

Theorem 15 *Let* L *be a formal language over* Σ *and let* $d : \Sigma^* \to \mathbb{N}^\infty$. *Then the following two statements are equivalent:*

(i) *There exists a context-free grammar with terminal alphabet* Σ *such that the number (possibly* ∞*) of distinct leftmost derivations of* w, $w \in \Sigma^*$, *from the start variable is given by* $d(w)$.

(ii) *There exists a pushdown automaton with input alphabet* Σ *such that the number (possibly* ∞*) of distinct computations from the initial instantaneous description for* w, $w \in \Sigma^*$, *to an accepting instantaneous description is given by* $d(w)$.

A pushdown automaton with input alphabet Σ is termed *unambiguous* iff, for each word $w \in \Sigma^*$ that is accepted, there exists a unique computation from the initial instantaneous description for w to some accepting instantaneous description.

Corollary 16 (Bucher, Maurer [4], Satz 2.3.30) *A formal language is generated by an unambiguous context-free grammar iff it is accepted by an unambiguous pushdown automaton.*

An additional result of this kind is also the following. (See Chomsky [5], Greibach [12], Kuich, Salomaa [24] and Corollary 5.12 of Kuich [22].)

Theorem 17 *Let $d : \Sigma^* \to \mathbb{N}$. Then the following three statements are equivalent:*

(i) *There exists a context-free grammar G with terminal alphabet Σ such that the number of distinct leftmost derivations of w, $w \in \Sigma^*$, from the start variable is given by $d(w)$.*

(ii) *There exists a context-free grammar G in the Chomsky normal form with terminal alphabet Σ such that the number of distinct leftmost derivations of w, $w \in \Sigma^+$, from the start variable is given by $d(w)$.*

(iii) *There exists a context-free grammar G in the Greibach normal form with terminal alphabet Σ such that the number of distinct leftmost derivations of w, $w \in \Sigma^+$, from the start variable is given by $d(w)$.*

Corollary 18 *The following four statements on an ε-free context-free language L are equivalent:*

(i) *There exists an unambiguous context-free grammar G such that $L(G) = L$.*

(ii) *There exists an unambiguous context-free grammar G in Chomsky normal form such that $L(G) = L$.*

(iii) *There exists an unambiguous context-free grammar G in Greibach normal form such that $L(G) = L$.*

(iv) *There exists an unambiguous pushdown automaton $\mathfrak{P}$ such that $||\mathfrak{P}|| = L$.*

5 Principal cones of algebraic power series

In this section, Σ_∞ denotes a fixed infinite alphabet. All finite alphabets Σ, possibly provided with indices, are subalphabets of Σ_∞.

A *rational representation* μ is a multiplicative morphism $\mu : \Sigma^* \to (A^{\mathrm{rat}}\langle\langle \Sigma'^* \rangle\rangle)^{Q \times Q}$. A *rational transducer* (*with input alphabet* Σ *and output alphabet* Σ')

$$\mathfrak{T} = (Q, \mu, S, P)$$

is given by

(i) a finite set Q of *states*,

(ii) a *rational representation* $\mu : \Sigma^* \to (A^{\mathrm{rat}}\langle\langle \Sigma'^* \rangle\rangle)^{Q\times Q}$,

(iii) $S \in (A^{\mathrm{rat}}\langle\langle \Sigma'^* \rangle\rangle)^{1\times Q}$, called *initial state vector,*

(iv) $P \in (A^{\mathrm{rat}}\langle\langle \Sigma'^* \rangle\rangle)^{Q\times 1}$, called *final state vector.*

The mapping $||\mathfrak{T}|| : A\langle\langle \Sigma^* \rangle\rangle \to A\langle\langle \Sigma'^* \rangle\rangle$ *realized* by a rational transducer $\mathfrak{T} = (Q, \mu, S, P)$ is defined by

$$||\mathfrak{T}||(r) = \sum_{q_1, q_2 \in Q} \sum_{w \in \Sigma^*} S_{q_1}(r, w)\mu(w)_{q_1,q_2} P_{q_2}, \quad r \in A\langle\langle \Sigma^* \rangle\rangle .$$

A mapping $\tau : A\langle\langle \Sigma^* \rangle\rangle \to A\langle\langle \Sigma'^* \rangle\rangle$ is called a *rational transduction* iff there exists a rational transducer $\mathfrak{T}$ such that $\tau(r) = ||\mathfrak{T}||(r)$ for all $r \in A\langle\langle \Sigma^* \rangle\rangle$. Our rational transductions are a straightforward generalization of the rational transductions of Berstel [2], Proposition III.7.3 to power series.

A rational transducer $\mathfrak{T} = (Q, \mu, S, P)$ specified as above can be considered to be a finite automaton equipped with an output device. In a state transition from state q_1 to state q_2, $\mathfrak{T}$ reads a letter $x \in \Sigma$ and outputs the rational power series $\mu(x)_{q_1,q_2}$. A sequence of state transitions outputs the product of the power series of the single state transitions. All sequences of length n of state transitions from state q_1 to state q_2 reading a word $w \in \Sigma^*$, $|w| = n$, output the power series $\mu(w)_{q_1,q_2}$. This output is multiplied with the correct components of the initial and the final state vector, and $S_{q_1}\mu(w)_{q_1,q_2}P_{q_2}$ is said to be the *translation* of w by transitions from q_1 to q_2. Summing up for all $q_1, q_2 \in Q$, $\sum_{q_1,q_2\in Q} S_{q_1}\mu(w)_{q_1,q_2}P_{q_2} = S\mu(w)P$ is said to be the *translation* of w by $\mathfrak{T}$. Multiplication by (r, w) and summation over all $w \in \Sigma^*$ yields then $||\mathfrak{T}||(r)$ for $r \in A\langle\langle \Sigma^* \rangle\rangle$.

Our basic semiring will now be $A\langle\langle \Sigma_\infty^* \rangle\rangle$. The subsemiring of $A\langle\langle \Sigma_\infty^* \rangle\rangle$ containing all power series whose supports are contained in some Σ^* is denoted by $A\{\{\Sigma_\infty^*\}\}$, i. e.,

$$A\{\{\Sigma_\infty^*\}\} = \{r \in A\langle\langle \Sigma_\infty^* \rangle\rangle \mid \text{there exists a finite alphabet } \Sigma \subset \Sigma_\infty \text{ such that } \mathrm{supp}(r) \subseteq \Sigma^*\}.$$

For $\Sigma \subset \Sigma_\infty$, $A\langle\langle \Sigma^* \rangle\rangle$ is isomorphic to a subsemiring of $A\{\{\Sigma_\infty^*\}\}$. Hence, we may assume that $A\langle\langle \Sigma^* \rangle\rangle \subset A\{\{\Sigma_\infty^*\}\}$. Furthermore, we define two subsemirings of $A\{\{\Sigma_\infty^*\}\}$, namely, the semiring of algebraic power series by

$$A^{\mathrm{alg}}\{\{\Sigma_\infty^*\}\} = \{r \in A\{\{\Sigma_\infty^*\}\} \mid \text{there exists a finite alphabet } \Sigma \subset \Sigma_\infty \text{ such that } r \in A^{\mathrm{alg}}\langle\langle \Sigma^* \rangle\rangle\},$$

and the semiring of rational power series by

$$A^{\mathrm{rat}}\{\{\Sigma_\infty^*\}\} = \{r \in A\{\{\Sigma_\infty^*\}\} \mid \text{there exists a finite alphabet } \Sigma \subset \Sigma_\infty \text{ such that } r \in A^{\mathrm{rat}}\langle\langle\Sigma^*\rangle\rangle\}.$$

Let $\mathfrak{T} = (Q, \mu, S, P)$ be a rational transducer with input alphabet Σ. The extended mapping $||\mathfrak{T}|| : A\{\{\Sigma_\infty^*\}\} \to A\{\{\Sigma_\infty^*\}\}$ is then defined by

$$||\mathfrak{T}||(r) = \sum_{q_1,q_2 \in Q} \sum_{w \in \Sigma^*} S_{q_1}(r, w)\mu(w)_{q_1,q_2} P_{q_2}, \quad r \in A\{\{\Sigma_\infty^*\}\}.$$

Here the rational representation μ is extended to $\mu : \Sigma_\infty^* \to (A^{\mathrm{rat}}\{\{\Sigma_\infty^*\}\})^{Q\times Q}$ by $\mu(w) = 0$ if $w \notin \Sigma^*$. Hence $\mathfrak{T}$ translates only words w of Σ^* and maps words not in Σ^* to 0. Again, a mapping $\tau : A\{\{\Sigma_\infty^*\}\} \to A\{\{\Sigma_\infty^*\}\}$ is called a *rational transduction* iff there exists a rational transducer $\mathfrak{T}$ such that $\tau(r) = ||\mathfrak{T}||(r)$ for all $r \in A\{\{\Sigma_\infty^*\}\}$.

Each nonempty subset $\mathfrak{L}$ of $A\{\{\Sigma_\infty^*\}\}$ is called *family of power series.* In the sequel, $\mathfrak{L}$ will denote a family of power series. A family of power series closed under rational transductions is called a *cone.* Define

$$\hat{\mathfrak{M}}(\mathfrak{L}) = \{\tau(r) \mid r \in \mathfrak{L} \text{ and } \tau \text{ is a rational transduction}\}.$$

(For $A = \mathbb{B}$, our operator $\hat{\mathfrak{M}}$ coincides with the operator $\hat{\mathcal{M}}$ of Ginsburg [8]; see also Nivat [27], Berstel [2].) Since rational transductions are closed under functional composition (Kuich [22], Theorem 7.9), $\mathfrak{L}$ is a cone iff $\mathfrak{L} = \hat{\mathfrak{M}}(\mathfrak{L})$. A cone $\mathfrak{L}$ is called *principal* iff there exists a power series $r \in A\{\{\Sigma_\infty^*\}\}$ such that $\mathfrak{L} = \hat{\mathfrak{M}}(r)$. The power series r is then a *cone generator* of the cone $\mathfrak{L}$. (See Berstel [2]; and Ginsburg [8], where cones are called full trios.)

We now want to characterize principal cones of algebraic power series by algebraic systems of a certain form. An algebraic system is in *cone form* iff it is of the form

$$y_i = \sum_{\pi \in Z} a_{i\pi} x_{i\pi} \pi, \quad 1 \le i \le n,$$

where $a_{i\pi} \in A$, $a_{iy_i} = 1$, $x_{i\pi} \in \Sigma$, $\pi \in Z$, $Z = Y^3 \cup Y^2 \cup Y \cup \{\varepsilon\}$, $1 \le i \le n$, and $\{x_{i\pi} \mid \pi \in Z - \{y_i\},\ 1 \le i \le n\} \cap \{x_{iy_i} \mid 1 \le i \le n\} = \emptyset$.

With respect to the generation of principal cones the cone form is a normal form.

Theorem 19 (Kuich [23]) *Each principal cone in $A^{\mathrm{alg}}\{\{\Sigma_\infty^*\}\}$ is generated by the first component of the unique solution of some algebraic system in cone form.*

Let $\mathfrak{S}$ be an algebraic system $y_i = p_i$, $1 \le i \le n$, in cone form, i. e.,

$$\begin{aligned} p_i = {} & (p_i, x_i)x_i + \textstyle\sum_{1\le j\le n}(p_i, x_{ij}y_j)x_{ij}y_j + \\ & \textstyle\sum_{1\le j,k\le n}(p_i, x_{ijk}y_jy_k)x_{ijk}y_jy_k + \\ & \textstyle\sum_{1\le j,k,m\le n}(p_i, x_{ijkm}y_jy_ky_m)x_{ijkm}y_jy_ky_m, \qquad 1 \le i \le n, \end{aligned}$$

where $x_i, x_{ij}, x_{ijk}, x_{ijkm} \in \Sigma$, $(p_i, x_{ii}y_i) = 1$, $1 \le i, j, k, m \le n$, and $(\{x_i \mid 1 \le i \le n\} \cup \{x_{ij} \mid 1 \le i, j \le n,\ j \ne i\} \cup \{x_{ijk} \mid 1 \le i, j, k \le n\} \cup \{x_{ijkm} \mid 1 \le i, j, k, m \le n\}) \cap \{x_{ii} \mid 1 \le i \le n\} = \emptyset$. Then, by definition, $\hat{\mathfrak{M}}(\mathfrak{S})$ is the collection of all the following algebraic systems written in matrix form:

$$\begin{aligned} Y_i = {} & (p_i, x_i) \otimes \mu(x_i) + \textstyle\sum_{1\le j\le n}(p_i, x_{ij}y_j) \otimes \mu(x_{ij})Y_j + \\ & \textstyle\sum_{1\le j,k\le n}(p_i, x_{ijk}y_jy_k) \otimes \mu(x_{ijk})Y_jY_k + \\ & \textstyle\sum_{1\le j,k,m\le n}(p_i, x_{ijkm}) \otimes \mu(_{ijkm})Y_jY_kY_m, \qquad 1 \le i \le n. \end{aligned}$$

Here $Y_i = (y^i_{jk})_{1\le j,k\le t}$, $1 \le i \le n$, is a $t \times t$-matrix of variables and $\mu : \Sigma^* \to (A\langle\Sigma' \cup \varepsilon\rangle)^{t\times t}$ is a rational representation, $t \ge 1$. Moreover, $\otimes$ denotes the Kronecker product (see Kuich, Salomaa [24]).

The y^1_{1t}-component of the least solution of the algebraic system defined above is called the *principal* component of this least solution. We are now ready to characterize principal cones in $A^{\mathrm{alg}}\{\{\Sigma^*_\infty\}\}$ by the principal components of the least solutions of the algebraic systems in $\hat{\mathfrak{M}}(\mathfrak{S})$.

Theorem 20 (Kuich [23]) *Let r be the first component of the unique solution of an algebraic system $\mathfrak{S}$ in the cone form. Then the principal cone $\hat{\mathfrak{M}}(r)$ coincides with the collection of all principal components of least solutions of the algebraic systems in $\hat{\mathfrak{M}}(\mathfrak{S})$.*

Corollary 21 *$\mathfrak{L} \subseteq A^{\mathrm{alg}}\{\{\Sigma^*_\infty\}\}$ is a principal cone iff $\mathfrak{L}$ coincides with the collection of all principal components of least solutions of the algebraic systems in $\hat{\mathfrak{M}}(\mathfrak{S})$ for some algebraic system $\mathfrak{S}$ in the cone form.*

For $A = \mathbb{B}$ this is a characterization of principal cones by context-free grammars.

Example. The *restricted one counter languages* form a principal cone that is generated by the Łukasiewicz language (see Greibach [13], Berstel [2]). These restricted one counter languages are generalized by Kuich, Salomaa [24] to power series. Again these *restricted one counter power series* form a principal cone and again it is generated by the solution of the algebraic system $y = x_1yy + x_2$. Clearly, this principal cone is also generated by the solution of the algebraic system $y = x_1yy + xy + x_2$ in cone form. Hence, a power series is a restricted one counter power series iff it is the y_{1t}-component of the least solution of an algebraic system $Y = M_1YY + MY + M_2$, $t \ge 1$. Here Y is a $t \times t$-matrix of variables y_{jk}, $1 \leq j,k \leq t$ and $M, M_1, M_2 \in (A\langle \Sigma \cup \varepsilon \rangle)^{t \times t}$, $t \geq 1$, $\Sigma \subset \Sigma_\infty$.

6 Decidability questions

In this last section we quote a few decidability results on formal languages and automata. We want to emphasize that no proofs in terms of classical automata and language theory exist for these decidability results. It seems that the technique of power series (and the resulting apparatus from classical mathematics) is quite essential for the proofs.

The first three results are originally due to Semenov [36]. (See also Salomaa, Soittola [33], Section IV.5 and Kuich, Salomaa [24], Section 16.)

Theorem 22 *Assume that G and G' are given unambiguous context-free grammars such that $L(G') \supseteq L(G)$. Then it is decidable whether or not $L(G') = L(G)$.*

Theorem 23 *Assume that G and G' are given context-free grammars such that $L(G') = L(G)$ and G is unambiguous. Then it is decidable whether or not G' is unambiguous.*

Theorem 24 *Given an unambiguous context-free grammar G and a regular language R, it is decidable whether or not $L(G) = R$.*

A context-free grammar $G = (Y, \Sigma, P, y_1)$ is termed *nonexpansive* iff there is no derivation $y \Rightarrow^* w_1 y w_2 y w_3$, $y \in Y$, $w_1, w_2, w_3 \in \Sigma^*$, according to G.

Theorem 25 (Ibarra, Ravikumar [17]) *Assume that G is a given unambiguous nonexpansive context-free grammar. If $L(G)$ is regular, a finite automaton $\mathfrak{A}$ can be effectively constructed such that $\|\mathfrak{A}\| = L(G)$.*

Ullian [40] has shown that an algorithm for the construction of $\mathfrak{A}$ does not exist, if G is an arbitrary context-free grammar. The algorithm for the construction of $\mathfrak{A}$ in Theorem 25 uses Theorem 24. Compare Theorem 24 and Theorem 25 with Stearns [39], Theorem 6.

The next theorem is based on an extension of the Equality Theorem of Eilenberg [7] to multitape finite automata.

Theorem 26 (Harju, Karhumäki [15], Theorem 3.10) *Given two multitape deterministic finite automata, it is decidable whether or not they accept the same language.*

Our last theorem states the solution to a long standing open problem posed by Ginsburg, Greibach [9].

Theorem 27 (Sénizergues [38]) *Given two deterministic pushdown automata, it is decidable whether or not they accept the same language.*
(For a full proof see Sénizergues [37].)

We hope that we could convince the reader that the results, constructions and proofs given in this paper satisfy the items (i)–(v) stated in the Introduction.

Acknowledgement. Thanks are due to Jean Berstel and Arto Salomaa for useful comments.

References

1. Berstel, J. (ed): Séries formelles en variables non commutatives et applications. Laboratoire d'Informatique Théorique et Programmation. Ecole Nationale Supérieure de Techniques Avancées, 1978.
2. Berstel, J.: Transductions and Context-Free Languages. Teubner, 1979.
3. Berstel, J., Reutenauer, C.: Les séries rationelles et leurs langages. Masson, 1984.
   English translation: Rational Series and Their Languages. EATCS Monographs on Theoretical Computer Science, Vol. 12. Springer, 1988.
4. Bucher, W., Maurer, H.: Theoretische Grundlagen der Programmiersprachen. B. I. Wissenschaftsverlag, 1984.
5. Chomsky, N.: On certain formal properties of grammars. Inf. Control 2(1959) 137–167.
6. Chomsky, N., Schützenberger, M. P.: The algebraic theory of context-free languages. In: P. Braffort and D. Hirschberg, eds., Computer Programming and Formal Systems. North-Holland, 1963, 118–161.
7. Eilenberg, S.: Automata, Languages and Machines. Vol. A. Academic Press, 1974.
8. Ginsburg, S.: Algebraic and Automata-Theoretic Properties of Formal Languages. North-Holland, 1975.
9. Ginsburg, S., Greibach, S.: Deterministic context-free languages. Inf. Control 9(1966) 620–648.
10. Ginsburg, S., Rice, H. G.: Two families of languages related to ALGOL. J. Assoc. Comput. Mach. 9(1962) 350–371.
11. Goldstern, M.: Vervollständigung von Halbringen. Diplomarbeit, Technische Universität Wien, 1985.
12. Greibach, S.: A new normal-form theorem for context-free phrase structure grammars. J. Assoc. Comput. Mach. 12(1965) 42–52.
13. Greibach, S.: An infinite hierarchy of context-free languages. J. Assoc. Comput. Mach. 16(1969) 91–106.
14. Harrison, M. A.: Introduction to Formal Language Theory. Addison-Wesley, 1978.
15. Harju, T., Karhumäki, J.: The equivalence problem of multitape finite automata. Theor. Comput. Sci. 78(1991) 347–355.
16. Hopcroft, J. E., Ullman, J. D.: Introduction to Automata Theory, Languages, and Computation. Addison-Wesley, 1979.
17. Ibarra, O. H., Ravikumar, B.: On sparseness, ambiguity and other decision problems for acceptors and transducers. STACS 86, Lect. Notes Comput. Sci. 210(1986) 171–179.

18. Karner, G.: On limits in complete semirings. Semigroup Forum 45(1992) 148–165.
19. Krob, D.: Monoides et semi-anneaux continus. Semigroup Forum 37(1988) 59–78.
20. Kleene, St. C.: Representation of events in nerve nets and finite automata. In: C. E. Shannon, J. McCarthy, eds., Automata Studies. Princeton University Press, 1956, 3–41.
21. Kuich, W.: The Kleene and the Parikh theorem in complete semirings. ICALP 87, Lect. Notes Comput. Sci. 267(1987) 212–225.
22. Kuich, W.: Semirings and formal power series: Their relevance to formal languages and automata theory. In: Handbook of Formal Languages (Eds.: G. Rozenberg and A. Salomaa), Springer, 1997, Vol. 1, Chapter 9, 609–677.
23. Kuich, W.: Cones, semi-AFPs and AFPs of algebraic power series. Technische Universität Wien, 2000.
24. Kuich, W., Salomaa, A.: Semirings, Automata, Languages. EATCS Monographs on Theoretical Computer Science, Vol. 5. Springer, 1986.
25. Manes, E. G., Arbib, M. A.: Algebraic Approaches to Program Semantics. Springer, 1986.
26. Moll, R. N., Arbib, M. A., Kfoury, A. J.: An Introduction to Formal Language Theory. Springer, 1988.
27. Nivat, M.: Transductions des langages de Chomsky. Ann. Inst. Fourier 18(1968) 339–455.
28. Paz, A.: Introduction to Probabilistic Automata. Academic Press, 1971.
29. Sakarovitch, J.: Kleene's theorem revisited. Lect. Notes Comput. Sci. 281(1987) 39–50.
30. Salomaa, A.: Theory of Automata. Pergamon Press, 1969.
31. Salomaa, A.: Formal Languages. Academic Press, 1973.
32. Salomaa, A.: Power from power series. MFCS 79, Lect. Notes Comput. Sci. 74, 170–181, 1979.
33. Salomaa, A., Soittola, M.: Automata-Theoretic Aspects of Formal Power Series. Springer, 1978.
34. Schützenberger, M. P.: Un problème de la théorie des automates. Seminaire Dubreil-Pisot, 13.3, Inst. H. Poincaré, 1960.
35. Schützenberger, M. P.: On the definition of a family of automata. Inf. Control 4(1961) 245–270.
36. Semenov, A. L.: Algoritmitšeskie problemy dlja stepennykh rjadov i kontekstnosvobodnykh grammatik. Dokl. Akad. Nauk SSSR 212(1973) 50–52.
37. Sénizergues, G.: $L(A) = L(B)$? Technical report, Nr. 1161–97, LaBRI,

Université Bordeaux I, 1967.

38. Sénizergues, G.: The equivalence problem for deterministic pushdown automata is decidable. ICALP 97, Lect. Notes Comput. Sci. 1256(1997) 671–681.
39. Stearns, R. E.: A regularity test for pushdown machines. Inf. Control 11(1967) 323–340.
40. Ullian, J. S.: Partial algorithm problems for context-free languages. Inf. Control 11(1967) 80–101.
41. Wechler, W.: The Concept of Fuzziness in Automata and Language Theory. Akademie-Verlag, 1978.

PLAYING INFINITE GAMES IN FINITE TIME

ROBERT MCNAUGHTON
Department of Computer Science
Rensselaer Polytechnic Institute
Troy, NY 12180-3590, U.S.A.
mcnaught@cs.rpi.edu

In the early 1960's Buchi and Muller, from different points of view, found themselves involved with finite automata having infinite histories. Their innovation inspired many new ideas in the decades that followed, one of which was a novel type of infinite game. In a recent paper [1] I have formulated this game as being played on a finite directed graph: two opposing players take turns forever in moving a token from node to node; winning and losing depends on that subset of nodes that are visited forever by the token.

It is well known that one of the two players has a winning strategy in every such game. In Section 7 of [1] I suggest a method of keeping score so that after some finite number of moves, a probable winner can be declared. The actual play of the game can be stopped at this point, whereupon it is clear that, if the two players were to play forever as well as they have been playing so far, the declared winner would be the true winner. In this way we can convert an infinite game on a finite graph to a finite game on the same graph. Depending on the size of the graph, it may then be humanly playable with the aid of a computer to keep score. This idea is explored in detail in this paper.

We believe that infinite games might have an interest for casual living-room recreation. Most games have an objective that one of the two players can usually achieve at some point in the play of the game; the moment of achievement is the moment that the play ends. Infinite games, on the other hand, go on forever. The play never ends, and so the outcome of the game cannot be defined by the situation at a final moment. If infinite games are interesting their interest lies in the fact that the players' objectives lie in a continuing sequence of states that one of the player achieves and is capable of perpetuating. For that reason infinite games are interesting in a way that games with a single final objective are not. They resemble real-life activity, in that winning is not something that happens in one final moment, but is something that one must sustain over an indefinite period of time.

Many of our life activities could serve as examples, but running a business seems most typical. Success, although it is not spread over an infinite duration is spread over an indefinite duration; it does not come in a single moment of consummation, but must be sustained over a long time interval. Along these lines Nerode et al in Section 3 of [2] consider in detail an infinite game that involves two kinds of move: adding to and deleting from a stack of goods. The objective is to move forever without ever making the stack too large, and without ever attempting to remove more goods than there are in the stack.

One can think of an infinite game as an idealization of such real-life activities: we represent something in life that lasts an indefinite length of time as something ideal that lasts an infinite length of time. In this paper we begin with the ideal concept of an infinite game, and eventually propose a method of justifiably stopping the play of a game at some moment and declaring one of the two players to be the winner. The justification of the declaration will be the judgment that, if the play were to continue with each playing forever as he has so far, then the player declared to be the winner would be the winner of the infinite play of the game.

In many recent papers on infinite games the intended application is the analysis of the running of concurrent computer programs. Since this paper is concerned with an entirely different kind of application, it will diverge in character from most of those papers.

The first section of this paper defines the class of infinite games that we consider. Section 2 reviews the score function, which was introduced in my previous paper [1]. Section 3 contains some results about this function, some from [1] and some new. Section 4 presents the method, using the score function, of deciding when the play of the game can be terminated and who should be declared the winner. Finally, Sections 5 and 6 offer a suggestion on how we might generate infinite games that human players may find interesting and possible to play.

1 Infinite games

In this paper a game $\mathcal{G}$ is an ordered quadruple (Q, E, W, Ω, n_0) where (Q, E) is a directed graph, Q being the set of nodes and E the set of directed edges; $\emptyset \neq W \subseteq Q$, $\Omega \subseteq 2^W$ and $n_0 \in Q$. We stipulate that, for each $p \in Q$, there is at least one $q \in Q$ such that $(p, q) \in E$, and at least one path from n_0 to p.

The game has two players Black and Red, who take turns moving a *token* from node to node along an edge. The play begins at node n_0 with Black moving it from there. The two players move in turn forever.

The set W is called the set of *winning-condition nodes.* For any play of the game, the *permset* of that play is the subset of nodes of W that are visited infinitely often by the token. If the permset is a member of Ω then Black wins; if it is a member of $2^W - \Omega$ then Red wins.

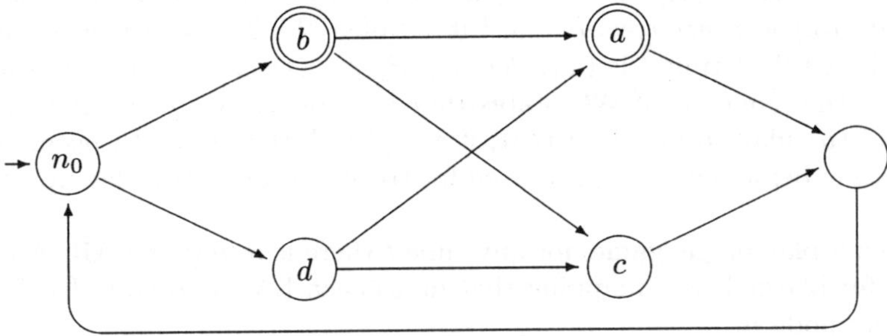

Figure 1. A game

As an example consider the game whose graph is shown in Figure 1. In this game $W = \{a, b\}$ and $\Omega = \{W, \emptyset\}$. Thus Black wins if and only if either both b and a are visited only finitely often or else they are each visited infinitely often; Red wins if and only if one of these two nodes is visited infinitely often and the other finitely often.

Black has a winning strategy for this game, which is as follows: from n_0 move to d if a has never been visited or if b has been visited since the last time a was visited; in all other circumstances move to b. If Black plays this strategy then Red can cause a and b to be visited only finitely often (e.g., by always moving from b or d to c) or he can cause them both to be visited infinitely often (e.g., by always moving to a). But, in playing against Black's strategy, Red cannot play so that one node of W is visited infinitely often while the other is visited only finitely often. So he cannot win. It is an important fact that Black has no winning strategy in this game that specifies moves without regard to the past. For example, if from n_0 Black always moves to b then Red can win by always moving from b to c. And if Black always moves to d from n_0 then Red can win by always moving from d to a.

In [1] it is proved that every game in the class of games considered there had a winning strategy for one of the two players of any game. As we shall see, the same is true of the class of games considered in this paper. To treat these strategies with precision we need an important concept:

The *state* of a game at any time t is (n, B) or (n, R), where node n is the location of the token at time t; if it is Black's turn to move from n at time t then the state is (n, B); if it is Red's turn then it is (n, R). If $w \in W$ then we say both (w, R) and (w, B) are *WC (winning-condition) states.* An LAR *(latest-appearance record* — in [1] it is called "LVR") is a sequence of WC states: the LAR at time t is $(w_1, P_1), (w_2, P_2), \ldots, (w_h, P_h)$, which is without repetition the sequence of WC states that have occurred up to time t in the history of the play so far, such that, if $j < j' \leq h$ then (w_j, P_j) occurred for the last time before $(w_{j'}, P_{j'})$ occurred for the last time in the play up to time t.

Given a play of the game, for any time t there is a unique LAR. An *LAR strategy* for player P is a mapping that maps each LAR and state (n, P) to a move from node n.

In recent papers about infinite games on graphs, the WC nodes of the game graph are assigned colors, each color being assigned generally to more than one node. Nodes of the same color are interchangeable as they affect winning and losing, which is defined in terms of the set of colors that occur infinitely often. The concept of coloring is ignored in most of this paper; another way to say this is to say that, in almost every game that we consider, no two WC nodes will be colored the same.

In perhaps all papers about infinite games on graphs except this one (but including [1]) graphs are bipartite: that is to say, the nodes are partitioned into the nodes from which Red can play and those from which Black can play, which means that it is not possible for one player to play from some node and then some time later the opposite player to play from that same node. Games whose graphs are not bipartite will be useful in this paper.

It is easy to construct, from a given graph G that is not bipartite, an equivalent bipartite graph G': Let each node n in G have two nodes n_B and n_R in G' (Black being the player who plays from n_B and Red the player who plays from n_R). G' is completed by drawing, for each directed edge (m, n) of G, directed edges (m'_B, n'_R) and (m'_R, n'_B). The initial node of G' is n_{0B}.

G' will have twice as many nodes as G and twice as many winning-condition nodes (although G' is sometimes simplifyable). The two WC nodes in the new graph replacing each WC node in the original graph will have the same color. Thus the winning condition for the game of G' in terms of colors will be exactly the same as the winning condition of the original game. The new game will be equivalent to the original game. Note that the nodes of the graph G' correspond one-to-one to the states of the original graph G.

Observe that the graph of Figure 1 is already bipartite, and so the construction is not necessary. The winning strategy for Black in that game, which is given above, is an LAR strategy. In this paper we shall find it convenient to use graphs that are generally not bipartite. For our purposes, there is no practical need to convert these graphs into bipartite graphs. On the other hand, for background theoretical understanding we shall refer to games whose graphs are bipartite.

This paper can be regarded as a continuation of Section 7 of my previous paper [1], which introduces the idea of playing an infinite game in finite time. From what is proved the earlier sections of [1] (which is based largely on the work of others), one of the two players of any game has a winning LAR strategy. Since LAR strategies are sufficient to offer solutions to all games, the player of an infinite game need not remember the entire history of the game in order to make an intelligent move, however long the play has been going; it suffices to keep track of the LAR of which there are finitely many. Indeed, we can say that there are only finitely many distinguishable configurations in an infinite graph game; the number is bounded by the number of states times the number of LAR's. Two configurations are indistinguishable if the state and the LAR are the same, because the move by the player is determined by the node and the LAR.

2 The score function

Our concern is infinite games that can be played meaningfully in a finite amount of time. Ultimately the goal (discussed but not achieved in this paper) is to have a game or games that are interesting to play and can be decided, perhaps by a referee, after being played a reasonable amount of time. The basic tool from Section 7 of [1] is the score function, which offers the opportunity of judging which subset α of W shows promise of being the permset, were the game to continue forever with both players playing forever as they have been playing up to that time. Roughly speaking, $score(\alpha, t)$ is high if the token has since some time $t' < t$ repeatedly and for many times visited all the nodes of α, while avoiding all the nodes of $W - \alpha$. We proceed to give a precise definition.

Let time 0, time 1, ... be the successive instants relevant to the game. At time 0 the token is at n_0; at time 0 Black moves the token to another node where it is at time 1; at time 1 Red moves it to another node where it is at time 2, etc. We define $node(t)$ to be the node visited by the token at time t. Thus $n_0 = node(0)$; $node(1)$ and $node(2)$ are, respectively, the node to which Black moves it and the node to which Red then moves it; etc.

For $\alpha \subseteq W$ and $node(t) \notin W - \alpha$, let $begin(\alpha, t)$ be the smallest nonnegative integer $t' \leq t$ such that

$$\{node(x) \cap W | t' \leq x \leq t\} \subseteq \alpha$$

Thus $begin(W,t) = 0$ for all t; and for $\alpha \neq W$, if $begin(\alpha,t) > 0$ then $node(begin(\alpha,t) - 1) \in W - \alpha$. The function $begin(\alpha,t)$ is undefined if $node(t) \in W - \alpha$. With these definitions the score function can be defined.

We first define $score(\alpha,t)$ for $\alpha \neq \emptyset$. For $node(t) \notin W - \alpha$ and $\alpha \neq \emptyset$, we define $score(\alpha,t)$ (the *score* for α at time t) to be the maximum nonnegative integer h such that the interval between time $begin(\alpha,t)$ and time t inclusive can be decomposed into h disjoint intevals $I_1,\ldots,I_h$ ($I_i < I_j$ for $i < j$) such that $node(I_i) \cap W = \alpha$ for each i; for $node(t) \in W - \alpha$, $score(\alpha,t) = 0$. (By *interval* we mean a set of consecutive integers. We write $node(I)$ to mean $\{node(t) | t \in I\}$.)

Note that if $node(t) \notin W - \alpha$ and $score(\alpha,t) = 0$ then for some $p \in \alpha$, either p was not visited at all by the token at time t or before, or else the token visited some $q \in W - \alpha$ since the last time it visited p. Also note that, for any $\alpha \neq \emptyset$ and t, if $node(t+1) \notin W - \alpha$ then $score(\alpha, t+1) =$ either $score(\alpha,t)$ or $score(\alpha,t) + 1$.

To complete this definition we need to define $score(\emptyset,t)$ (which was left undefined in [1]). Let $Q - W = \{f_1,\ldots,f_s\}$. For each i, $1 \leq i \leq s$, we define

$$score_i(\emptyset,0) = \begin{cases} 1 \text{ if } node(0) = f_i \\ 0 \text{ otherwise} \end{cases}$$

$$score_i(\emptyset,t+1) = \begin{cases} 0 & \text{if } node(t+1) \in W \\ score_i(\emptyset,t) & \text{if } node(t+1) = f_j \text{ and } j \neq i \\ score_i(\emptyset,t) + 1 & \text{if } node(t+1) = f_i \end{cases}$$

Note that $score_i(\emptyset,t)$ is the number of times the token has visited the node f_i since the last time it visited a node of W, or since the beginning of the play if it has never visited a node of W.

From the functions $score_1,\ldots score_s$, we define

$$score(\emptyset,t) = \max_{1 \leq i \leq s} score_i(\emptyset,t)$$

Loosely speaking, if $node(t) \notin W$, then $score(\emptyset,t)$ is one more than the number of times the token loops on some fixed node of $Q - W$ during the interval from $begin(\emptyset,t)$ to t. It does not matter if some of the nodes of $Q - W$ are not visited at all.

In any infinite play and for any integer h, however large, there exists a t and an $\alpha \subseteq W$ such that $score(\alpha,t) = h$. In particular, this will be true when α is the permset of the play. But it is also possible for this to be true when α is not the permset.

With the score function we can approach the task of formulating a justifiable method of determining when one of the two players can be declared a winner. The idea is that a player wins when one of his sets α gets a sufficiently high score. The question is, how high must the score of α be before the play can be terminated? For the present let us simply note that, ideally, termination should not occur unless at least one of the following two conditions obtains:

1. since $begin(\alpha, t)$ the losing player never had a winning LAR strategy to play; or

2. for every winning LAR strategy the loser had the opportunity of playing in the interval from $begin(\alpha, t)$ to t, there was some time in the interval when he could have moved according to that strategy but failed to do so.

In section 4 we shall discuss the practical aspects of deciding when to terminate the play and declare a winner. For now we note merely that the computation of the score function is an indispensable part of our plan.

In order to make the game playable, I suggest a computer program to do a running computation of $score(\alpha, t)$ for every $\alpha \subseteq W$ for the active duration of the play of the game, and to display all positive values of the score function for the players. The method of keeping track of $score(\emptyset, t)$ is apparent from the definition as given. The method of keeping track of $score(\alpha, t)$ for $\alpha \neq \emptyset$ is more complicated:

Consider the history of a game to time t where $score(\alpha, t) > 0$ and let t_0 be the last time before t when $node(t_0) \in W - \alpha$. Clearly, $score(\alpha, t_0) = 0$. The first time t_1 after t_0 when $score(\alpha, t_1) = 1$ is the earliest time when $node(t_1) \in \alpha$ and, for all $n \in \alpha$ other than $node(t_1)$, $node(t') = n$ for some t', $t_0 < t' < t$. For any t'', $t_0 < t'' < t$, we define $subs(\alpha, t'')$ to be the subset of α of nodes that have occurred since t_0 up to and including time t''.

Similarly, let t_2 be the earliest time when $score(\alpha, t_2) = 2$. For any t''' between t_1 and t_2, $subs(\alpha, t''')$ is the subset of the nodes of α that have occurred since t_1 up to and including time t'''. Etc.

Formally, we define the subs function recursively:

$$subs(\alpha, 0) = \begin{cases} \{node(0)\} & \text{if } node(0) \in \alpha \\ \emptyset & \text{otherwise} \end{cases}$$

$$subs(\alpha, t+1) = \begin{cases} \emptyset & \text{if } node(t+1) \in W - \alpha \\ subs'(\alpha, t) \cup \{node(t+1)\} & \text{if } node(t+1) \in \alpha \\ subs'(\alpha, t) & \text{otherwise} \end{cases}$$

where

$$subs'(\alpha, t) = \begin{cases} \emptyset & \text{if } subs(\alpha, t) = \alpha \\ subs(\alpha, t) & \text{otherwise} \end{cases}$$

We now have a way of computing $score(\alpha, t)$ recursively, for any $\alpha \neq \emptyset$:

$$score(\alpha, 0) = \begin{cases} 1 \text{ if } \alpha = \{node(0)\} \\ 0 \text{ otherwise} \end{cases}$$

$$score(\alpha, t+1) = \begin{cases} score(\alpha, t) + 1 & \text{if } subs(\alpha, t+1) = \alpha \\ 0 & \text{if } node(t+1) \in W - \alpha \\ score(\alpha, t) & \text{otherwise} \end{cases}$$

We leave it to the reader to verify that the function $score(\alpha, t)$ computed by these equations is identical to the function as originally defined.

The above equations demonstrate that the score function is easily computed by a computer during the play of the game. In any game that we can expect people to want to play, the number of subsets of W that the players have to be cognizant of could not be too large; otherwise the game would be too difficult for the players to focus on. Either the set W would have to be reasonably small, or else only a limited number of subsets of W could be possible permsets. For this reason, I believe that an ordinary personal computer could do the service of keeping track of the score function for the players, and for the referee.

3 Theoretical results about scoring

The possibility of playing an infinite game in finite time was established by the results in [1], which we summarize now, along with some needed extensions. Some of these results are valid only for games with bipartite graphs; for various reasons it does not serve our purpose to extend these results to games whose graphs are not bipartite, even though the latter will be the games that we shall ultimately consider.

A *memoryless game* is a game in which the winning strategies need not refer to LAR's; in other words they are strategies in which how a player is directed to play from any node does not refer to the past history of the token. A player following a memoryless strategy will move the same way from a node every time he moves from that node. In [1] memoryless games are discussed at length. Of course, not all games are memoryless games; e.g., the game of Figure 1 is not a memoryless game, as we have noted.

In a memoryless game U_B (U_R) is the set of all nodes from which Black (Red) has a winning memoryless strategy. Thus $U_B \cup U_R$ is the set of all nodes of the graph and $U_B \cap U_R = \emptyset$.

Theorem 3.1 (Theorem 7.2 of [1]). Let $\mathcal{G}$ be a memoryless game whose graph is bipartite. Suppose $t' > t$ and, in some play of $\mathcal{G}$, $node(t') = node(t)$ and

$$\{node(t'')|t' \leq t'' \leq t\} \cap W = \alpha \in 2^W - \Omega.$$

Then either no node of U_B is visited between t' and t or else, for every winning memoryless strategy f for Black, at least one node of U_B is visited some time when Black played but not according to f.

Theorem 3.1 and other theorems of this section are stated for $\alpha \in 2^W - \Omega$. By symmetry, these results are equally valid for $\alpha \in \Omega$ and the rôles of Red and Black reversed.

We note that if, $\alpha \in 2^W - \Omega$, $node(t) \in \alpha$ and $score(\alpha, t) = 2$ in a memoryless game whose graph is bipartite, then the antecedent, and hence the consequent, of Theorem 3.1 are satisfied. It follows that, in this circumstance, the referee may be justified in declaring Red the winner. The reason is that Red has succeeded in getting the token to go from a node n of α to visit all the other nodes of α and then back to node n without visiting any nodes of $W - \alpha$. (In fact this may have happened before the score of α becomes 2. But it cannot happen until some time after the score becomes 1.)

Note the importance of the condition in Theorem 3.1 that the game be a memoryless game. Had the game not been a memoryless game the referee might not be justified in declaring Red the winner. Black could have been playing a winning LAR strategy that, although it permitted Red to loop through α once, would foil Red's attempt to loop through again upon a change of LAR's.

A class of memoryless games that has received attention recently from many authors is the class of *parity games* (see, e.g., [4]), i.e., those games in which some nodes are colored from the set $\{c_1, c_2, \cdots, c_h\}$ (h a positive integer) and where winning and losing depends on whether the maximum j for which c_j occurs infinitely often is odd or even. My present thinking is that interesting games for us will be those requiring strategies involving memory. However, if some day a contrary thought comes to the fore, parity games might be worth considering in devising infinite games for people to play.

In a game that is not a memoryless game, U_B (U_R) is the set of all nodes from which Black (Red) has a winning LAR strategy. In spite of the change in meaning, it is still true that $U_B \cup U_R$ is the set of all nodes of the graph and $U_B \cap U_R = \emptyset$.

Theorem 3.2 (Theorem 7.3 of [1]). Let $\mathcal{G}$ be a game whose graph is bipartite, and let $\emptyset \neq \alpha \in 2^W - \Omega$ where $|\alpha| = k$. If $score(\alpha, t) = k! + 1$ then either (a), for all t', $begin(\alpha, t) \leq t' \leq t$ implies $node(t') \notin U_B$, or (b), for any winning LAR strategy f for Black, there exists a t', $begin(\alpha, t) \leq t' < t$, such that $node(t') \in U_B$ and Black played but not according to f at time t'.

In this circumstance the referee would not be justified in declaring Red the winner until Red is able to loop through α a sufficient number of times to make sure that Black is not playing an LAR strategy that would eventually foil Red's attempt to keep looping through α.

Although proofs are not given here for Theorems 3.1 and 3.2 (the reader is referred to [1]) we should explain the significance of the required score of $k! + 1$ in Theorem 3.2, and why no such score is required in Theorem 3.1. In Theorem 3.2 the required score must be large enough to guarantee that if Black were playing an LAR strategy that could prevent the permset from being α, the score of α could not be that high. The required score must be high enough to guarantee that some node from which Black plays is repeated with the same LAR. Since the play during that time does not visit any node of $W - \alpha$, but circulates through all of α, the LAR's assumed during that period is from a set of $k!$ LAR's. If the score for α reaches $k! + 1$ then some node must be visited twice with the same LAR.

In Theorem 3,1, on the other hand, the game is assumed to be a memoryless game, and so a player moving from the same node more than once will make the same move each time. Thus the required score for each set α can be taken as 2 (although the referee for such a game might insist on a higher score greater than this to guard against chance results due to player inexperience in the early stages of the play).

Memoryless games would have the advantage of a justified earlier termination. However, memoryless games are not as interesting as those games whose winning strategies require players to move differently from the same node at different times. What we should look for are games that are not memoryless games but where the winning strategies are such that the number of different moves dictated by differing LAR's is small.

We might be aided here by a recent deep result by Dziembowski, Jurdziński and Walukiewicz [3]; see also [4]. They define the memory required by a strategy for a player in an infinite game to be the minimum number of states of a finite automaton that can play for that player according to that strategy. Their result is a fairly simple method of computing the memory required by a given player of any game in the class of games having a common winning condition. In other words, they give an upper bound on the memory requirement for strategies in all games with a specified winning condition.

The significance of this result in the topic of this paper is that it could give us a way of finding games that are not memoryless games but whose strategic memory requirements are much less than the memory requirements for all games. However, the pragmatic approach put forth in the later sections of this paper will not make use of this idea.

Instead, we shall introduce another theoretical concept. Assume a game and define two LAR's to be *equivalent* relative to a pair (σ_B, σ_R) of LAR strategies for Black and Red, respectively, if each player, moving according to the appropriate strategy from any node, would move the same with response to the two LAR's. Of interest is the *index* of (σ_B, σ_R), by which we mean the *index* of this equivalence relation, i.e., the number of equivalence classes.

Consider the class of all such pairs (σ_B, σ_R) where σ_B (σ_R) is a winning strategy for Black (Red) playing from any node of U_B (U_R). Let (τ_B, τ_R) be a member of this class of pairs whose index is minimal in the class. Where $\mathcal{G}$ is the game under consideration let $m_{\mathcal{G}}$ be the *index* of (τ_B, τ_R).

The index of a memoryless game is 1. The index of every game is $\leq |W|!$. I suspect that the index of most games is much smaller then $|W|!$. At any rate, it would be in our interest to find a class of games whose indices exceed 1 but not be very much. The justification of this remark is in the following refinement of Theorem 3.2, whose proof (i.e., the proof of Theorem 7.3 in [1]) is easily adapted:

Theorem 3.3. Let $\mathcal{G}$ be a game whose graph is bipartite and let $\emptyset \neq \alpha \in 2^W - \Omega$ and $|\alpha| = k$. If $score(\alpha, t) = \min(k!, m_{\mathcal{G}}) + 1$ then either (a), for all t', $begin(\alpha, t) \leq t' \leq t$ implies $node(t') \notin U_B$, or (b), for any winning LAR strategy f for Black, there exists a t', $begin(\alpha, t) \leq t' < t$, such that $node(t') \in U_B$ and Black played but not according to f at time t'.

Theorem 3.4 (Theorem 7.4 of [1]). Let $\mathcal{G}$ be a game, and let $\emptyset \in 2^W - \Omega$. If $t' < t$ and, during a play of $\mathcal{G}$, $node(t') = node(t)$ and $\{node(t'') | t' \leq t'' \leq t\} \cap W = \emptyset$ then either no node of U_B is visited between t' and t or else, for every winning LAR strategy f for Black, at least one node of U_B is visited some time when Black played but not according to f.

As long as the token avoids nodes of W, the LAR does not change. Thus during such a play of the game LAR's are immaterial to the winning strategy for Red, and immaterial for any strategy for Black that could force the token to a node of W.

4 When to end the play

The two theorems of this section, unlike those of the preceding section, apply to games whose graphs are not necessarily bipartite.

Theorem 4.1 (Theorem 7.1 of [1]). For $t \geq 2$ and $\alpha, \beta \subseteq W$, if

$$score(\alpha, t) > score(\alpha, t-1) > 0 \text{ and } score(\beta, t) > score(\beta, t-1) > 0$$

then $\alpha = \beta$.

In other words , if $\alpha \neq \beta$ then the scores for both α and β cannot both increase to 2 or more from time $t-1$ to time t. Note that it is possible for the scores of distinct α and β to simultaneously increase if one of them, say α increases from 0 to 1. This can happen only if $\alpha \subset \beta$. E.g., suppose $\alpha = \{a, b\}$, $\beta = \{a, b, c\}$, $node(0) = c$, $node(1) = b$, $node(2) = a$, $node(3) = c$, $node(4) = b$, and $node(5) = a$. Then $score(\alpha, 4) = 0$, $score(\alpha, 5) = 1$, $score(\beta, 4) = 1$ and $score(\beta, 5) = 2$. In general, if $\alpha \subset \beta$ it is possible for $score(\beta, t-1)$ to be arbitrarily large, $score(\beta, t) = score(\alpha, t-1) + 1$, $score(\alpha, t-1) = 0$ and $score(\alpha, t) = 1$.

This observation shows that whatever policy we adopt for declaring a winner in finite time, we should never do so by virtue of the score of any set reaching the value of 1. Another reason for this policy is that a score of 2 shows that there has been a loop of the set, while a score of 1 in itself does not.

Theorem 4.1 shows that whatever specific policy we institute following our general method for declaring a winner, provided it is in accord with the principle just formulated, there will never be a time when we would be obliged to declare both players to be the winner, since the first time any set will reach its desired score cannot be the first time any other set will reach its desired score.

Theorem 4.2. In any game, if $\alpha, \beta \subseteq W$ and $\alpha - \beta \neq \emptyset \neq \beta - \alpha$ then, for all t, either $score(\alpha, t) = 0$ or $score(\beta, t) = 0$.

Proof: Assume our theorem is false, i.e., $score(\alpha, t) > 0$, $score(\beta, t) > 0$ and there exist nodes n_1 and n_2 such that $n_1 \in \alpha - \beta$ and $n_2 \in \beta - \alpha$. Let t_1 be the last time before time t that $node(t_1) = n_1$, and let t_2 be the last time before time t that $node(t_2) = n_2$. We get

$$\begin{aligned} t_1 &< begin(\beta, t) && \text{(since } n_1 \notin \beta) \\ begin(\beta, t) &\leq t_2 && \text{(since } n_2 \in \beta) \\ t_2 &< begin(\alpha, t) && \text{(since } n_2 \notin \alpha) \\ begin(\alpha, t) &\leq t_1 && \text{(since } n_1 \in \alpha) \end{aligned}$$

We thus get $t_1 < t_1$, a contradiction.

By Theorem 4.2, if $\alpha \neq \beta$, $score(\alpha, t) \neq 0$ and $score(\beta, t) \neq 0$ then either $\alpha \subset \beta$ or $\beta \subset \alpha$. It should be noted that if, say, $\alpha \subset \beta$ it is possible during a game for $score(\alpha)$ and $score(\beta)$ to be large simultaneously. In that case, where $t' = begin(\alpha, t)$, it must be that $begin(\beta, t) < t'$, $score(\beta, t') =$ either $score(\beta, t)$ or $score(\beta, t) - 1$, and the score of α has increased from t' to t with the score of β holding steady most of the time since $time(t')$. (If $score(\beta, t') = score(\beta, t) - 1$ then the score of β increases to $score(\beta, t)$ when the score of α becomes 1 or sooner.)

In short, with $\alpha \subset \beta$, if $score(\alpha, t)$ and $score(\beta, t)$ are both large then β achieved at least most of its score before the score of α began to increase.

There are two possible game situations here, depending on whether α and β are winning sets for the same player or for opposite players. If α and β are winning sets for the same player, say Black, then this might be a situation where she is playing a winning strategy; but which of her sets actually wins for her depends on how Red, in losing, plays his losing game. This situation is exactly that of the game of Figure 1 in Section 1 with $\alpha = \emptyset$ and $\beta = W$. (As in [1] it is assumed, for the sake of pronomial convenience, that Black is female and Red is male. In much of the literature on games my Black is called "player 0" and my Red is called "player 1".)

On the other hand, if α and β are winning sets for opposite players the situation may be disturbing and interesting. If α is a winning set for Red and β a winning set for Black, at time t' it looks as if Black is winning. But then Red starts to threaten by confining the play to α, during which time the score for α increases while the score for β remains steady after at most one increase after time t'. At time t if Red can continue to confine the play to α he will win; but if Black manages to regain control and enlarge the play again to all of β then she will win.

To see how disturbing and interesting this situation might be, suppose the score of β goes up to just 2 points below where Black would be declared the winner and Red suddenly sees how he can confine the play to α. In this situation Black will lose the game but with quite a high score for her set β. The decision to award the game to Red is not incorrect, since Red did ultimately find a strategy that succeeded in beating Black's strategy.

A big part of our task will be that of finding games that are most interesting to play. One observation in this connection is that memoryless games are not too interesting. Infinite games are more interesting when the winning player must vary his moves made at the same node at different times. Thus we cannot use Theorem 3.1 exclusively in formulating which score values permit terminating the play of a game. Generally speaking, for interesting games most sets will require a score significantly greater than 2 to terminate the play.

We therefore consider Theorem 3.2, which tells us that an adequate score for α to terminate the play in a game whose graph is bipartite is $k! + 1$, where $k = |\alpha|$. If the graph is not bipartite the adequate score may be much larger. As we shall argue in Section 6, the games of interest would not be bipartite and would have $|W| \geq 5$. For $|W| = 5$, if we were to rely only on what Theorem 3.3 tells us then a play of a game in which W is the permset would require a score of more than $5! + 1 = 121$, in fact probably much more if the graph is not bipartite. This would involve an unacceptably long time.

For this practical reason the referee's policy about when to end the play of the game cannot be based simply on the theorems that have so far been proved. This policy will have to be adopted pragmatically, according to the following guidelines:

(1) The referee must arrange to have the score of all subsets of W available throughout the play of the game (probably by a computer). Also, this information should be available to the players.

(2) Each such α should have a *cut-off score* c_α such that if

$$score(\alpha, t) = c_\alpha = score(\alpha, t-1) + 1$$

then the play ends at time t with Black the winner if $\alpha \in \Omega$, and Red the winner if $\alpha \in 2^W - \Omega$.

(3) The c_α's should be chosen perhaps by judgment based on experience with games actually played. The concept of the index $m_\mathcal{G}$ of a game $\mathcal{G}$ (see Section 3) is relevant here, although an easy way of computing $m_\mathcal{G}$ for a given $\mathcal{G}$ is not known.

(4) For $|\beta| > |\alpha|$, c_β should be greater than c_α. For $|\beta| = |\alpha|$, c_β and c_α should be equal.

(5) Each c_α should be large enough so that if $score(\alpha, t) = c_\alpha$ then the play can be terminated at time t without the losing player feeling unjustly treated.

(6) Each c_α should be small enough so that plays of the game are not too long.

(7) The c_α's should be large enough so that there is a good chance that a play of the game will end by consent of the loser before the score of any α reaches c_α.

In considering the problem of when to end the play of the game it should be kept in mind that our plan is to find a subclass of these games which can be handled by referees and enjoyed by players, neither of whom need solve the game mathematically. Possibly we shall want the refereeing done by computer.

This and the previous two sections have been concerned with the score function and finding a justifiable procedure of deciding when to terminate an infinite game and declare a winner. This is a difficult and vital problem. However appealing a game is to players, it would not serve our purpose if the players must play an excessive amount of time for an outcome; potential players would soon lose interest in our games. Likewise if the play of a game is halted prematurely, without a clear indication that the player who is declared the winner would win the infinite game (given that both players play forever as they have played up to the time of the declaration) then that would also cause potential players to lost interest.

It seems reasonable to adopt a compromise between these two extremes. That is to say, declaring the winner after it is reasonably certain that the declared winner has played substantially better than the declared loser, but not absolutely certain.

5 Finding playable games

Although the problem of deciding when to end the play of an infinite game is difficult and important, it is not the most vital of our problems. More crucial to our endeavor are problems having to do with human factors. We must make sure, for example, that the game arouses and sustains the interest of potential players. In this section we shall discuss our plans to construct games with this objective in mind.

To begin, there is the task of setting up the game so that it is playable move by move. The graph must be realized physically so that (1) the position of the token and the alternative moves are clear, and (2) each player's objective is clear and meaningful enough to enable a reasonably quick move.

It should be noted that if the set of winning-condition nodes is large and the set Ω (the set of winning permsets for Black) is complex, the players would never get to the point where they could do anything but make random moves and leave things to chance. On the other hand, if W is not too large and Ω is, for example, the class of subsets of W having even cardinality then the game might be quite playable. Black would play to attempt to make the permset even, and Red would play to attempt to make it odd. (The game of Figure 1 is an example of this idea — much too simple, of course, for our present purpose.)

Another idea would be to make $|W| = 5$ and take Ω to be the class of subsets of W with cardinality three, four or five. In playing this game Black would attempt to make the cardinality of the permset at least 3, and Red would attempt to make it at most 2.

It is a variation on this last idea that is our favored suggestion: that the objective for Black shall be to make the permset as large as possible and for Red to to make it as small as possible. This game design would go beyond the binary concept of winning and losing, which is adhered to in most of the literature, in [1], and in the early sections of this paper. But it seems optimal, among the alternatives, for making the players' objectives natural and meaningful. We shall now adopt this idea tentatively, without being certain that it will work out; for the remainder of this paper we shall not discuss alternatives.

Theorem 5.1. For every infinite game, there is an $n \leq |W|$, and strategies σ_B for Black and σ_R for Red such that playing σ_B will guarantee Black that the size of the permset will be at least n and playing σ_R will guarantee Red that the size of the permset will be at most n.

This theorem follows from Theorem 4.1 of [1]. Since the graphs here are not necessarily bipartite and the payoff function replaces the set Ω, the inference is not obvious. However, it is not difficult.

Using the results of [3] and [4] it can be proved that σ_R in Theorem 5.1 can be a memoryless strategy. Thus, if this game is actually played by two people without knowledge of the value of n in Theorem 5.1, the person playing Red could restrict himself to a style of play in which his move would depend only on the current position of the token, without regard to the history of the play. The person playing Black, on the other hand, should consider varying her play at certain nodes depending on the past history of the game. Thus we might expect this game to be more challenging for Black than for Red.

Theorem 5.1 tells us that optimal strategies of any game are effectively determined. But in order not to spoil the fun, we should be sure that the players not know these strategies and that they not know the value of n of Theorem 5.1. The plan suggested here is that neither the players nor the referee should be capable of, or even desirous of, doing the calculation needed to figure out these strategies or the value of n. Rather, we should expect that the players will learn about how to play the game informally and intuitively in the course of actually playing it. We cannot expect that, under these circumstances, either player will succeed in learning an optimal strategy.

We must design games that are not too hard and not too easy. They must be neither transparent nor incomprehensible to the players. Either extreme would diminish the players' pleasure of participation.

If we succeed in making the game enjoyable, many players will want to play again. If they play exactly the same game over and over, i.e., the game with exactly the same graph and set W, they would eventually learn the particularities of the game so as to know it almost completely, whereupon it would no longer be challenging enough to be enjoyable. To avoid this possibility, we should arrange to have many games (each given by a graph and set W) available, so that a person would not have to play the same game twice. The games would differ from one another, although they would all be based on the same principles. A player who plays again would be like a bridge player playing another hand for which the cards are shuffled.

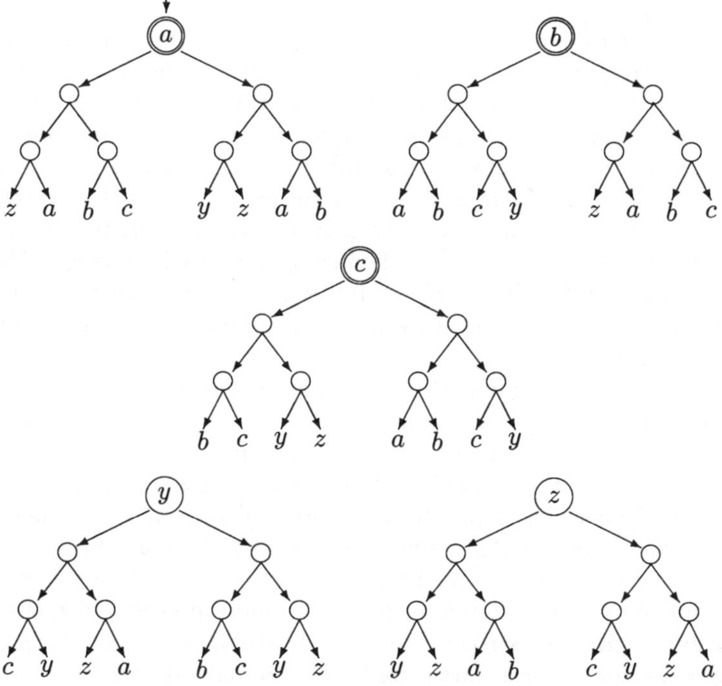

Figure 2. A game for reference

6 A specific suggestion

A specific procedure will now be suggested for finding games that we hope interested people can play some day. All graphs will be strongly connected but not bipartite. In all our graphs, every node that one player moves from may at some future time be the place where the opposite player moves from. This property is desirable because it ensures that the total number of states is twice the number of nodes. (Recall from Section 1 that the state in which the token is at node N with Black about to move is different from the state in which the token is at N when it is Red's move.) The graphs that we use will be considerably more compact than the equivalent bipartite graphs, which would generally require twice as many nodes. For that reason they offer us a human-factors advantage: the graph that is presented to the player is simpler and easier to look at than the equivalent bipartite graph would be.

We now present an example of a game, given by Figure 2. The figure consists of five trees, but represents a strongly connected directed graph with vertices and arcs as shown, except that each arc at the bottom of each tree pointing to a letter signifies that the arc in the graph points to the vertex whose label is that letter. The node labeled a is the initial node and $W = \{a, b, c\}$. The winning condition for the game of this graph is as described in Section 5: Black seeks to make the permset large, and Red seeks to make it small. It turns out that Red has a memoryless strategy that forces the permset to be either $\{a\}$ or $\emptyset$. Playing against Red's strategy, Black can avoid a permset of $\emptyset$. The same would still be true if the initial node were any other node of the graph. (We leave the proof of these assertions to the industrious reader.)

The fact that this game is so transparent shows that the games to be played should be substantially larger.

Our plan is to have, for some integers m, n and p, $m+n$ symmetric binary directed trees with roots labeled, respectively, $a_1, a_2, \ldots, a_m, z_1, z_2, \ldots, z_n$. Each tree will have depth p, and therefore will have $2^{p+1} - 1$ nodes. (The value of p in Figure 2 is 2.) From each tree leaf there are two arcs pointing to tree roots (as in Figure 2). Exactly which trees these arcs point to will be determined by some random or pseudo-random process, with the restraint that the entire graph be strongly connected. We set $W = \{a_1, \ldots, a_m\}$. Black seeks to make the permset large; Red seeks to make it small.

What remains to be decided are: (1) Exactly what are to be the values of m, n, p? (2) Exactly what are to be the restraints on the labeling of the leaves of the original forest? The plan is, once these issues are settled, to generate many games as explained in Section 5. It seems that the values of m, n, p should be determined by intuition and experiment, without further theoretical investigation. Offhand I would say that m should be somewhere between 5 and 10 with n somewhere between 1 and $m/2$. My feeling is that, whatever the values of m and n, the value of p should be 2 for ease of playing. As an example, we might have $m = 7$, $n = 3$ and $p = 2$. The graph would then be like Figure 2 except that the original forest would have ten trees instead of five. Since each tree has only four leaves, no tree could be connected from below to more than eight trees. Unlike Figure 2, the graph would have the desirable property that no tree could be connected to all other trees; desirable because it would tend to make the game more challenging for the players.

The value of n must not be too large, otherwise games would last too long. Yet n should be at least 1 to relieve the monotony of the play. When m and p are decided on, it might be a good idea to experiment to find the optimal value of n.

As mentioned in Section 5, we can expect that each game will be more challenging for Black than for Red, since Red need consider only memoryless strategies. For that reason, people playing these games will probably want to alternate playing the rôles of Black and Red.

We should probably not want to figure out the mathematical solutions to the games that we generate (i.e., the winning strategies for the game and what they accomplish for their respective players). There need not be anyone who even cares to have this information.

References

1. R. McNaughton, "Infinite games played on finite graphs,' *An. Pure and Applied Logic,* vol. 65 (1993), pp. 149–184.
2. A. Nerode, J.B. Remmel and A. Yakhnis, "McNaughton games and extracting strategies for concurrent programs." *An. Pure and Applied Logic,* vol. 78 (1996), pp. 203–242.
3. S. Dziembowski, M. Jurdziński and I. Walukiewicz, "How much memory is needed to win infinite games?" *Proc.* 12th *An. IEEE Symposium on LICS* (1997), pp. 99–110.
4. W. Zielonka, "Infinite games on finitely coloured graphs with applications to automata on infinite trees." *Theoretical Computer Science,* vol. 200 (1998), pp. 135–183.

GENE ASSEMBLY IN CILIATES: COMPUTING BY FOLDING AND RECOMBINATION

G. ROZENBERG

Leiden Institute of Advanced Computer Science (LIACS)
Leiden Center for Natural Computing (LCNC)
Leiden University
Niels Bohrweg 1
2333 CA Leiden
The Netherlands
and
Department of Computer Science
University of Colorado at Boulder
Boulder, CO 80309, USA
E-mail: rozenber@liacs.nl

Introduction.

The interaction of theoretical computer science and biology dates back to the very beginning of computer science. The paper [20] by S.C. Kleene, one of the "founding papers" of automata theory, was inspired by the work of W.S. McCulloh and W. Pitts, see, e.g., [25], which considers neurons as binary transmitters of information. Also, the very influential book [27] by J. von Neuman was inspired by the functioning of living organisms.

The development of the theory of L systems, introduced by A. Lindenmayer in [24], is a prime example of a very fruitful interaction between biology and theoretical computer science; more specifically, between developmental biology and formal languages and automata theory. L systems had a fundamental impact in formal language theory as well as significant impact in modeling of plants (and computer graphics), see, e.g., [37] and [38].

The "DNA revolution", which in the last 50+ years had such tremendous impact on biology and many other areas of science (and on our everyday life), also had a big influence on theoretical computer science. So, e.g., the overwhelming success of Celera Genomics and the Human Genome Project in the sequencing of the human and other genomes, was to the large extent based on the development of pattern matching and other string processing algorithms, see, e.g., [26]. The whole area of pattern matching and editing algorithms, see, e.g., [2] and [6], benefited enormously from the very intensive research concerned with the sequencing of genomes. As a matter of fact, the design and analysis of string processing algorithms lies at the very hart of bioinformatics, see, e.g., [31].

Since DNA molecules have a very natural representation as (double) strings over the alphabet $\mathcal{N} = \{A, C, G, T\}$ of nucleotides, and *in vivo* or *in vitro* processing of DNA molecules can be seen as applying operations on strings, the connection between DNA processing and formal language theory is very natural.

This connection was used by T. Head, who in his pioneering work [15], proposed an elegant model for processing of DNA by restriction enzymes. This model called *splicing systems* or *H-systems* (in honor of T. Head) was introduced in 1987, a full seven years before the seminal paper by L. Adleman [1], that initiated the area of *DNA computing*, a fast growing, fascinating, and genuinely interdisciplinary research area, see, e.g., [4] and [30]. By now, the theory of splicing systems is a well developed research area which lies at the intersection of formal language theory and DNA computing (see, e.g., [16]).

The rapid growth of DNA computing has stimulated novel and interesting research in formal languages and automata theory (besides the theory of splicing systems). Examples of this research are *sticker systems* (see, e.g., [30]), *Watson-Crick automata* (see, e.g., [30]), and *membrane systems* (see, e.g., [28] and [29]).

It is also interesting to note that the universal computing power of many models of DNA computing has its roots in the theory of *twin shuffle languages* developed long before the launch of the DNA computing area – see, e.g., [14], [30] and [40].

Research in DNA computing can be divided roughly into two, certainly not disjoint, streams: DNA computing *in vitro* and DNA computing *in vivo.* The former is concerned with (the theoretical foundations of and experimental work on) building DNA based computers in the test tubes. DNA computing *in vivo* is concerned with constructing computational components (such as, e.g., simple switching circuits) in living cells, see, e.g., [21] and [42], and with studying computational processes taking place in living cells. The current paper presents research falling in this latter category: studying computational processes in living cells. In particular we discuss the computational nature of a very intricate DNA processing taking place in single cell organisms called *ciliates* during the process of *gene assembly.*

The area of DNA computing studies the use of biomolecules for the purpose of computing (*in vitro* or *in vivo*). Here, molecular biology assists computer

scientists to achieve a really bold goal: to replace (or to complement) the silicon based computers by the DNA based computers. On the other hand, in *bioinformatics* and in *computational biology* computer scientists and mathematicians assist molecular biologists in understanding the structure and functioning of biomolecules, such as DNA and proteins, in living cells. More generally one can say that bioinformatics and computational biology are concerned with the computational nature of complex biological phenomena.
Therefore, the research presented here lies at the intersection of all three areas: DNA computing, bioinformatics and computational biology. Clearly, this research fits perfectly into the area of *Natural Computing*, the term that I coined many years ago to denote the computing inspired by (gleaned from) nature.

The computational beauty of gene assembly in (hypotrichous) ciliates was brought to the attention of the DNA computing community in a series of papers by L. Landweber and L. Kari, see, e.g., [22] and [23] (see also [41]). They have proposed a computational model based on *intermolecular* operations postulated to accomplish gene assembly in ciliates. Their research concerns computational power of these operations – e.g., they have proved that their operations are Turing-universal.

A different line of research concerned with gene assembly in (hypotrichous) ciliates was initiated in [13]. The basic model considered here is based on a different set of molecular operations, and more important, it is *intramolecular*. Moreover, this line of research is concerned with the process of gene assembly itself (rather than with the problems of computational power in the sense of computability theory). The usefulness of this model has been demonstrated by applying it to *all* known cases of gene assembly in ciliates, see [33], obtaining in this way a uniform explanation of all cases. The present paper discusses, primarily, the results obtained in this line of research.

Research on computational aspects of gene assembly in ciliates turned out to be interesting both from the biology and the theoretical computer science point of view. One of the main contributions of this research to theoretical computer science is the better understanding of the novel computational paradigm: *computing by folding and recombination.* This paradigm underlies the present paper, which is written in a tutorial/survey style. Occasionally, we may use a biological term not explained in this paper (e.g., "miosis") but then such a term belongs to the vocabulary of the high school biology, and its meaning can be easily found in any dictionary of biology or genetics.

1 Preliminaries.

In this section we recall a number of notions concerning strings and graphs, mainly to establish the notation for this paper.

The empty set is denoted by $\emptyset$.
For an alphabet Σ (we consider only finite alphabets), Σ^* denotes the set of all strings over Σ including the empty string Λ (we consider only finite strings). For strings $\alpha, \beta \in \Sigma^*$, we say that β is a *substring* of α, if $\alpha = \gamma\beta\delta$ for some strings $\gamma, \delta \in \Sigma^*$. If $\alpha = x_1 ... x_n \in \Sigma^*$, with $x_1, ..., x_n \in \Sigma$, $n \geq 1$, then the substrings $x_{i_1} ... x_{j_1}$ and $x_{i_2} ... x_{j_2}$ of α *overlap* (*in* α) if $i_1 < i_2 \leq j_1$ and $j_1 < j_2$. We say that β is a *cyclic substring* of α, if either β is a substring of α or $\beta = \beta_1\beta_2$ and $\alpha = \beta_2\gamma\beta_1$ for some strings β_1, β_2, and γ.
We say that β is a *scattered substring* of α, if $\beta = \beta_1 ... \beta_n$ and $\alpha = \alpha_1\beta_1\alpha_2\beta_2 ... \alpha_n\beta_n\alpha_{n+1}$ for some $n \geq 1$, and some strings $\alpha_1, ..., \alpha_{n+1}, \beta_1, ..., \beta_n$. If $\beta \neq \Lambda$ is a substring (cyclic substring, scattered substring) of α and $|\beta| \neq |\alpha|$, then we say that β is a *proper substring (proper cyclic substring, proper scattered substring*, resp.) of α.

A *graph* $\gamma = (V, E)$ consists of a set V of *nodes*, and a set $E \subseteq \{\{x, y\} : x, y \in V \text{ and } x \neq y\}$ of (undirected) *edges*. For $\{x, y\} \in E$ we say that the nodes x, y are *adjacent* in γ. Then, for a node $x \in V$, the *neighborhood of* x (*in* γ), denoted by $\mathbf{ngb}_\gamma(x)$, consists of all nodes in V adjacent to x; hence $\mathbf{ngb}_\gamma(x) = \{y \in V : \{x, y\} \in E\}$. We say that $x \in V$ is *isolated* (*in* γ) if $\mathbf{ngb}_\gamma(x) = \emptyset$. If $V = \emptyset$, then γ is the *empty graph*, also denoted by $\emptyset$.

A *signed graph* $\gamma = (V, E, \theta)$ consists of a graph (V, E) and a *sign function* $\theta : V \to \{+, -\}$. A node $x \in V$ is *positive in* γ, if $\theta(x) = +$, and x is *negative in* γ, if $\theta(x) = -$.

The following construction on signed graphs will be used in Section 7.
Let $S \subseteq V$ be a subset of the set of nodes of a signed graph $\gamma = (V, E, \theta)$. The graph $\gamma' = (V, E', \theta')$ is obtained from γ by *complementing* S, if for all pairs $\{x, y\}$ of V with $x \neq y$, we have $\{x, y\} \in E'$ if and only if
- $x \notin S$ or $y \notin S$ and $\{x, y\} \in E$,
- $x \in S$ and $y \in S$ and $\{x, y\} \notin E$,

and moreover $\theta'(y) = -\theta(y)$ for all $y \in S$, and $\theta'(y) = \theta(y)$ if $y \notin S$.
If for a node $x \in V$ we complement $\mathbf{ngb}_\gamma(x)$, then we get the *local complementation at* x (*in* γ), denoted by $\mathbf{lcom}_x(\gamma)$.

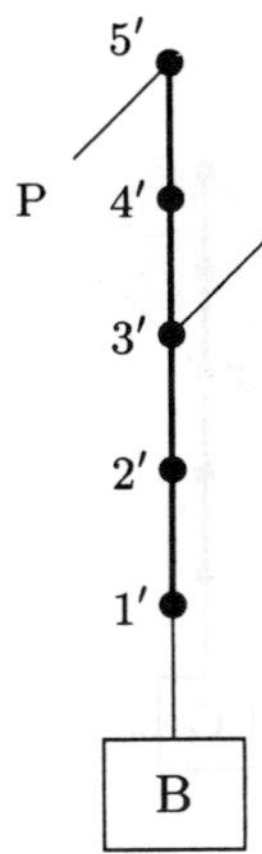

Figure 1. The structure of a nucleotide

2 DNA molecules: structure and notation.

DNA molecules are polymers built from simple monomers called *nucleotides* (DNA stands for "deoxyribonucleic acid"). A nucleotide consists of three components: *sugar*, *phosphate*, and *base*. The sugar molecule has five attachment points (carbon atoms) labeled $1'$ through $5'$, with the phosphate molecule attached to $5'$, and the base attached to $1'$. This is illustrated in Figure 1, where the sugar molecule is represented by a stick with the attachment points given by five fat dots, P denotes the phosphate molecule, and the square labeled B denotes the base.

There are four possible bases: A, C, G, and T (standing for Adenine, Cytosine, Guanine, and Thymine, respectively). The most important feature of the set of four bases is that they have a pairwise affinity: A with T, and C with G, which means that A and T, and C and G "like to stick together" (this will be clarified in the sequel). This pairwise affinity is called the *Watson-Crick*

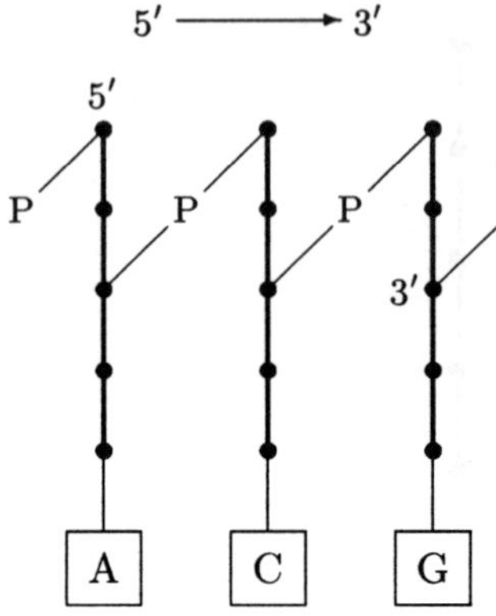

Figure 2. Single stranded DNA

complementarity. Nucleotides may differ only in their bases - therefore there are only four types of nucleotides, which are also denoted by A, C, G, and T.

Single stranded DNA molecules are just chains of nucleotides which are strung together by strong (covalent) bonds that form between the $3'$-attachment point of one nucleotide and the phosphate group on the $5'$-attachment point of the "next" nucleotide, see Figure 2. This covalent bond is called a *phosphodiester bond.* Note that one end of a single stranded DNA molecule has the $5'$-attachment point available for binding with yet another nucleotide, while the other end has the $3'$-attachment point available for binding. Since these attachment points can (easily) be distinguished chemically, a single stranded DNA molecule has a *polarity* (orientation) – we can read its nucleotide sequence either in the $5'$-$3'$ direction or in the $3'$-$5'$ direction. Since biological information is encoded in the $5'$-$3'$ direction, the nucleotide sequence is read in this way. Thus, e.g., for the single stranded DNA molecule from Figure 2, its nucleotide sequence reads as ACG.
A double stranded DNA molecule is formed by two single stranded DNA molecules that bind together through their bases by weak bonds (called *hydrogen bonds*). Two bases (one from one single stranded DNA molecule, and one from the other single stranded DNA molecule) can form such a bond only if these two bases constitute a complementary pair, i.e., either $\{A,T\}$ or $\{C,G\}$. Moreover, a double stranded DNA molecule can be formed only if

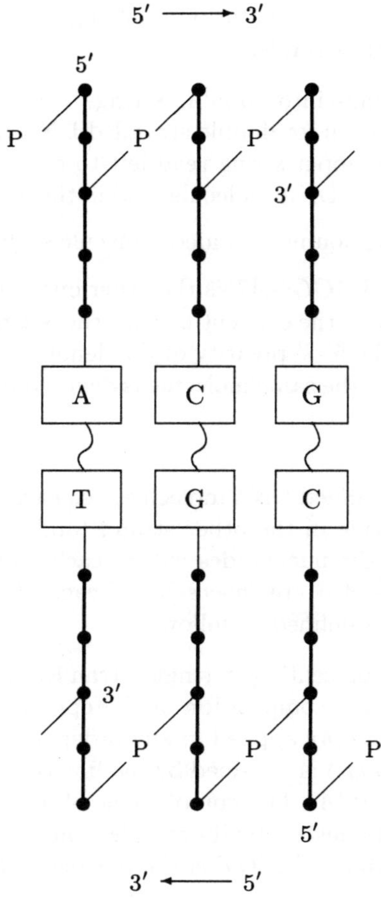

Figure 3. Double stranded DNA

the two single stranded DNA molecules are of opposite orientation - i.e., the first nucleotide at the 5′-end of one strand binds to the first nucleotide at the 3′-end of the other strand; the second nucleotide from the 5′-end of the first strand binds to the second nucleotide from the 3′-end of the second strand, etc. – see Figure 3.

Since we have four types of nucleotides, single stranded DNA molecules can be denoted by strings over the alphabet $\mathcal{N} = \{A, C, G, T\}$. Here we use the convention that the left-to-right orientation of strings over $\mathcal{N}$ corresponds to the $5'$-$3'$ orientation of single strands.

Now, combining single strings into "double strings", written one above the other, we get a natural notation for double stranded DNA molecules. Here we use the convention that the upper string read left-to-right denotes one of the strands of the double stranded DNA molecule read in the $5'$-$3'$ direction. Thus, e.g., $\frac{ATCCGT}{TAGGCA}$ denotes the double stranded molecule such that $ATCCGT$ is one of its single strands, and $ACGGAT$ is the other one (we wrote $ACGGAT$ and *not* $TAGGCA$, because of the convention that the left-to-right orientation of a string corresponds to the $5'$-$3'$-polarity of the denoted strand). Note that, according to our convention, the same molecule is denoted by the double string $\frac{ACGGAT}{TGCCTA}$.

Let α be a perfect double stranded DNA molecule, i.e., each nucleotide in α has the complementary nucleotide in the other strand (and with phosphodiester bonds present between all the nucleotides within each strand). We say then that the two single strands of α are *inversions* of each other. More formally the operation of inversion is defined as follows.

For a string β over $\mathcal{N}$ (representing a single stranded DNA molecule) its inversion $\bar{\beta}$ is obtained by the composition of the operations of *reversal* and (Watson-Crick) *complementation* applied in any order to β. The reversal of β (often called also *mirror image*) is obtained by reading β backwards – e.g., the reversal of $ACATG$ is $GTACA$. The complement of a string over $\mathcal{N}$ results by replacing each nucleotide (letter) by its complement – e.g., for $GTACA$ its complement is $CATGT$. Thus, $CATGT$ is the inversion of $ACATG$.

For a double string γ over $\mathcal{N}$ (representing a double stranded DNA molecule) its inversion $\bar{\gamma}$ is obtained by the composition of operations of reversal and *exchange*. Again, the reversal of γ results from reading γ backwards – e.g., the reversal of $\frac{ACATG}{TGTAC}$ is $\frac{GTACA}{CATGT}$. The exchange of a double string results by exchanging its two single strings for each other – e.g., the exchange for $\frac{GTACA}{CATGT}$ yields $\frac{CATGT}{GTACA}$. Thus, $\frac{CATGT}{GTACA}$ is the inversion of $\frac{ACATG}{TGTAC}$. As another example, consider the double string with "overhangs" $\begin{array}{l}CATGT\\ \quad ACAGG\end{array}$.

Its inversion is the double string with "overhangs" $\begin{array}{l}GGACA\\ \quad TGTAC\end{array}$.

Such overhangs are also called "sticky ends" because if two complementary sticky ends come together, then they bind together by establishing hydrogen bonds. For example, the molecule (denoted by the double string) $\begin{matrix} ATCTGAG \\ TAGA \end{matrix}$ and the molecule $\begin{matrix} CAATG \\ CTCGTTAC \end{matrix}$ have complementary sticky ends: $5'$-GAG-$3'$ in the first molecule, and $3'$-CTC-$5'$ in the second (here we have given explicitly the polarities of the sticky ends). Now, when these two molecules find each other, then their sticky ends will bind together. Hydrogen bonds are not strong, and the joining of the two molecules by hydrogen bonds only between GAG and CTC is rather tenuous. The molecule so formed, $\begin{matrix} ATCTGAGCAATG \\ TAGACTCGTTAC \end{matrix}$, has one missing covalent (phosphodiester) bond in each strand. However, if the enzyme *ligase* is present in the solution (and it is present in ciliates), then these missing bonds are established, and the two strands are now stably held together by many hydrogen bonds. This sticky end mechanism is central to the recombination process that will underlie our considerations of gene assembly. The (homologous) *recombination* process of the sort considered in this paper (viz., *intramolecular*) consists of
(i) the alignment of two (almost) identical segments of a given DNA molecule,
(ii) cutting the two segments in a staggered fashion so that sticky ends are formed, and
(iii) "exchanging" the parts of the segments using the complementary sticky ends.
Examples of such a recombination are given in Figures 7, 9, and 10.

The above presentation of the structure of, and the notation for DNA molecules is based on Chapter 1 from [30]. For more biological background, see, e.g., [3].

3 Gene assembly in ciliates.

Ciliates are a very ancient group of organisms – about two billions years old. This group contains more than 10000 genetically different organisms. They are single cell organisms that have developed a unique way of organizing their DNA. All eukaryotes keep their DNA protected in a nucleus, but ciliates have refined this system very significantly. They in fact have *two* kinds of nuclei, which are of different functionality: a germline nucleus (*micronucleus*) used in cell mating, and a somatic nucleus (*macronucleus*) providing RNA transcripts to operate the cell. As a matter of fact, both micronucleus and

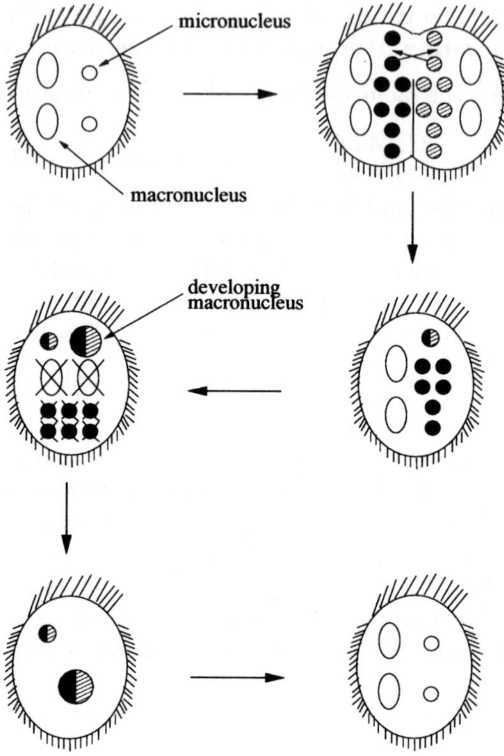

Figure 4. The mating scheme.

macronucleus are present in multiple copies, and the number of copies may be very different in different species. For example, *Oxytricha nova* has 2 copies of the micronucleus and 2 copies of the macronucleus, *Urostyla grandis* has 8–15 copies of micronucleus and 200–300 copies of macronucleus, and *Holosticha monilata* has 3–5 copies of micronucleus and 8–10 copies of macronucleus. However, the multiplicity of nuclei is irrelevant for our purposes in this paper.

The micronucleus is used for storing DNA until it is needed in the process of *sexual reproduction*. When starved, ciliates proceed to sexual reproduction (or they form cysts, or ... they die), which begins by *mating* – two cells unite temporarily through a cytoplasmic bridge, micronuclei undergo meiosis, and single haploid micronuclei are exchanged through the bridge. Then, an exchanged haploid micronucleus *fuses* with a stationary micronucleus to form a diploid micronucleus. The new diploid micronucleus *divides* by mitosis. One of the daughter micronuclei develops into a new macronucleus, the other daughter remains as a micronucleus. Simultaneously, the unused haploid micronuclei and the old macronuclei are degraded. Then the newly developed macronucleus and the micronucleus divide into two, producing a mature vegetative cell, thus completing the process of sexual reproduction. This is illustrated in Figure 4 (which is a modification of Figure 2 in [32]).

It is during the formation of the macronucleus (when a new diploid micronucleus develops into a macronucleus) that the micronuclear genome is converted into the macronuclear genome – this process is referred to as *gene assembly.* The gene assembly process is very intricate, because the micronuclear genome is *drastically different* from the macronuclear genome. The DNA in the micronucleus is hundreds of thousands of base pairs long, with genes occurring individually or in groups, and separated by stretches of noncoding spacer DNA (which accounts for about 95 % of the micronuclear DNA) – this is quite typical for a eukaryotic genome. On the other hand, the DNA molecules in macronucleus are gene-size, on average about 2000 base pairs long. Also, the form of the micronuclear and the macronuclear version of the same gene can be drastically different.

Micronuclear genes are interrupted by multiple noncoding segments called IESs (*Internal Eliminated Segments*). Gene segments separated by IESs are called MDSs (*Macronuclear Destined Segments*) – the IES/MDS structure of a micronuclear gene is as given in Figure 5. Here, each MDS M_i has the structure $M_i = \left(\frac{p_i}{\bar{p}_i}, \mu_i, \frac{p_{i+1}}{\bar{p}_{i+1}}\right)$ except for M_1 and M_k which are of the form $M_1 = \left(\frac{b}{\bar{b}}, \mu_1, \frac{p_2}{\bar{p}_2}\right)$, $M_k = \left(\frac{p_k}{\bar{p}_k}, \mu_k, \frac{e}{\bar{e}}\right)$, where b designates "beginning" and e designates "end". We refer to each $\frac{p_i}{\bar{p}_i}$ as a *pointer*, and to μ_i as the *body* of M_i. Each pointer occurs in an MDS at its boundary; $\frac{p_i}{\bar{p}_i}$ is called the *incoming pointer* of M_i, and $\frac{p_{i+1}}{\bar{p}_{i+1}}$ is called the *outgoing pointer* of M_i. Thus, each pointer has two occurrences: $\frac{p_i}{\bar{p}_i}$ is the incoming pointer of M_i and the outgoing pointer of M_{i-1}. Each p_i is in fact a sequence of nucleotides, and $\bar{p}_i$ is in the inversion of p_i. On the other hand, b and e (and $\bar{b}, \bar{e}$) are (symbolic) *markers* - we use them to simply mark the locations where an incipient macronuclear gene will be excised from the micronuclear DNA.

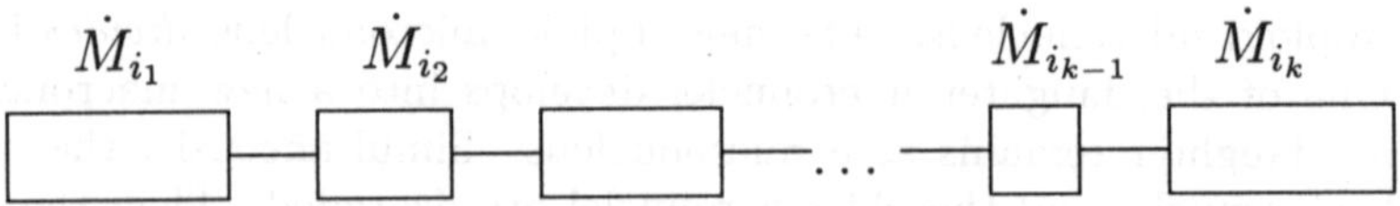

Figure 5. The structure of a micronuclear gene.

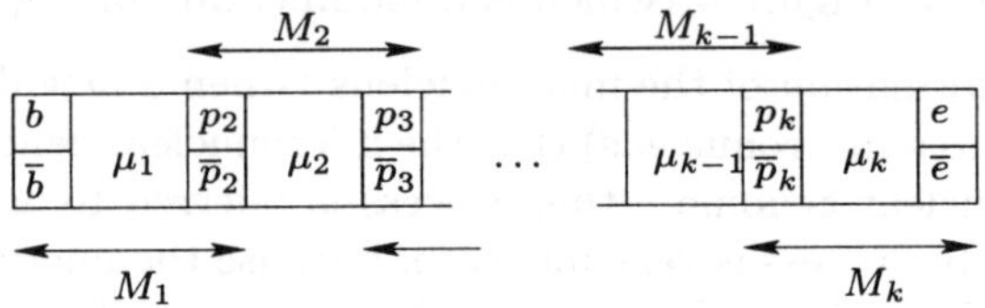

Figure 6. The structure of a macronuclear assembled gene.

The basic relationship between the micronuclear gene and the macronuclear gene is as follows: the macronuclear gene is obtained from its micronuclear form by splicing the MDSs $M_1, M_2, ...M_k$ in the *orthodox order* in the way shown in Figure 6 – each MDS $M_j, 1 \leq j \leq k$, is "glued" with M_{j+1} on the "overlapping" pointer $\frac{p_{i+1}}{\overline{p}_{i+1}}$ (the dot over M_{i_j} in Figure 5 means that this can be either M_{i_j} or its inversion).
Thus the sequence of MDSs forming the micronuclear gene (interspersed by IESs) is a permutation of the orthodox order, with possibly some of the MDSs inverted. For example, the micronuclear gene can be of the form $M_7 I_1 M_4 I_2 \overline{M}_5 I_3 M_1 I_4 M_2 I_5 \overline{M}_6 I_6 M_3$, where $I_1, ..., I_6$ are interspersing IESs.

As another example, the actin I gene in *Oxytricha nova* species of ciliates, see [35], is of the form $M_3 I_1 M_4 I_2 M_6 I_3 M_5 I_4 M_7 I_5 M_9 I_6 \overline{M}_2 I_7 M_1 I_8 M_8$, where the pointers $p_1, ..., p_8$ are the following sequences of nucleotides:
$p_1 = CTTACTACACAT$,
$p_2 = ATGTGTAGTAAG$,
$p_3 = AATC$,
$p_4 = CTCCCAAGTCCAT$,
$p_5 = GCCAGCCCC$,
$p_6 = CAAAACTCTA$,
$p_7 = CTTTGGGTTGA$,
$p_8 = AGGTTGAATGA$.

During the gene assembly process the MDSs of the micronuclear gene will be spliced together in the orthodox order, while the interspersing IESs will be excised.

Before we describe and formalize the gene assembly process, we will simplify our notation for pointers by using positive integers $2, 3, ..., k$, and $\bar{2}, \bar{3}, ..., \bar{k}$ to denote them and their inversions (and we keep $b, e, \bar{b}, \bar{e}$ to denote the markers and their inversions). Then we set $\Pi_k = \{2, 3, ..., k, \bar{2}, \bar{3}, ..., \bar{k}\}$. As a matter of fact, we will refer to *all* elements of Π_k as pointers, and for any pointer $p \in \Pi_k$ or marker $p \in \{b, e, \bar{b}, \bar{e}\}$, $\bar{p}$ denotes its inversion where $\bar{\bar{p}} = p$ (recall that markers b and e are just symbols marking the location where a macronuclear gene will be excised – hence, their "inversions" $\bar{b}$ and $\bar{e}$ are again just symbols). For any $p \in \Pi_k$, we call $\{p, \bar{p}\}$ the *pointer set* of p, and we denote it by $\mathbf{pts}(p)$ (and by $\mathbf{pts}(\bar{p})$). Using this simplified notation we get $M_1 = (b, \mu_1, 2)$, $M_k = (k, \mu_k, e)$, and $M_i = (i, \mu_i, i + 1)$, for all $2 \leq i \leq i - 1$. Then, for an MDS $M = (p, \mu, q)$, its inversion is $\bar{M} = (\bar{q}, \bar{\mu}, \bar{p})$. Also, a micronuclear gene τ, or an intermediate form τ of a macronuclear gene, has an *occurrence* of a pointer p, if p is either the incoming or the outgoing pointer of an MDS in τ.

In order to simplify the notation, we will fix k in the sense that we assume that, unless clear otherwise, we consider genes that have k MDSs in their micronuclear form.

We would like to make the following important comment. The term "gene assembly" means the process of transforming the micronuclear version in such a way that the MDSs become spliced together in their orthodox order, and the interspersing IESs are excised. This is a simplification which is adequate for the purpose of this paper, but one has to realize that the process of forming a macronuclear gene does not end here. The (sub)molecule consisting of MDSs spliced together has to be excised (at the locations indicated by markers b and e), and the telomer caps put on the ends of the excised molecule.

The reader is referred to [32], [33] and [34] for nicely written expositions of the biology of gene assembly. The notation used here for the MDS structure and pointers (markers) is from [13] and [11].

4 Molecular operations for gene assembly.

It is postulated in [13] and [36] that the gene assembly process is accomplished through applications of three molecular operations:

1. (*loop, direct repeat*)-*excision* (*ld-excision*, or just *ld* for short),
2. (*hairpin, inverted repeat*)-*excision/reinsertion* (*hi-excision/reinsertion*, or just *hi*, for short), and

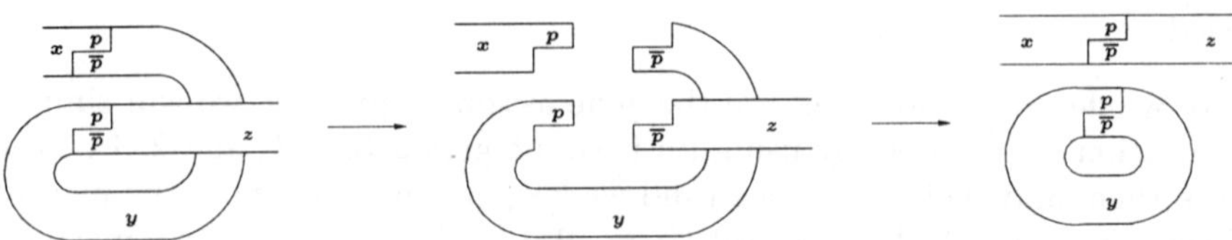

Figure 7. The *ld*-excision operation.

3. (*double loop, alternating direct repeat*)-*excision/reinsertion* (*dlad-excision/reinsertion*, or just *dlad* for short).

These operations are defined as follows.

1. The operation of *ld*-excision is applicable to molecules having a *direct repeat pattern* (p, p) of a pointer p, i.e., molecules that have two occurrences of pointer p. Such a molecule is folded into a loop in such a way that the two occurrences of p are aligned, and then the operation proceeds as shown in Figure 7. The excision here involves staggered cuts (producing sticky ends); therefore the operation yields one circular and one linear molecule. If an application of **ld** is a step in a successful gene assembly, then either (i) the two occurrences of pointer p are separated by one IES, and so the circular molecule consists of one IES only ("polluted" by one copy of pointer p), while in the linear molecule a bigger composite MDS is formed – see Section 8 for more discussion concerning this case, or (ii) all the MDSs and IESs (present before the operation is applied) end up in the circular molecule – this is the case when the two occurrences of a pointer p are on the two opposite ends of the molecule (before the operation is applied).
The second case is referred to as the *boundary* application of *ld*-excision, because the two occurrences of p are at the boundaries of the molecule – this is illustrated in Figure 8. As is easily seen (from the description of the two other molecular operations, given below), the two boundary occurrences of p may be affected only by the boundary application of *ld* – thus, in reasoning about successful gene assembly, one may always assume that this application is the very last operation.

2. The operation of *hi*-excision/reinsertion is applicable to molecules that have an *inverted repeat pattern* $(p, \bar{p})$ of a pointer p, i.e., molecules that have an occurrence of pointer p followed by an occurrence of $\bar{p}$. Such a molecule is folded into a hairpin in such a way that the occurrence of p and the occurrence of $\bar{p}$ are aligned, and then the operation proceeds as shown in Figure 9. The excision involves staggered cuts; then through reinsertion, the operation yields

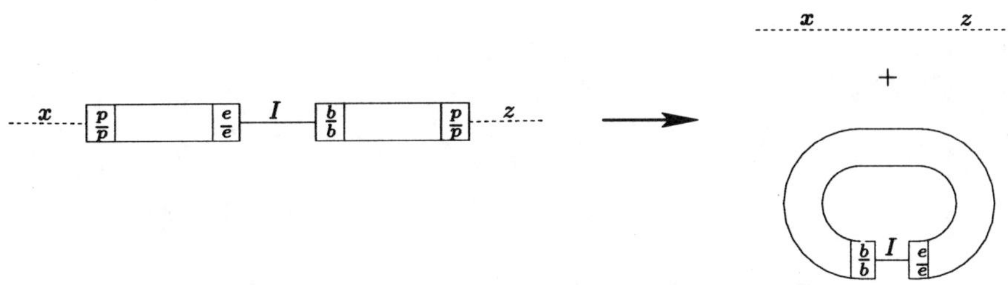

Figure 8. The boundary application of the *ld*-excision to the pointer 3 results in the assembled gene placed into a circular molecule.

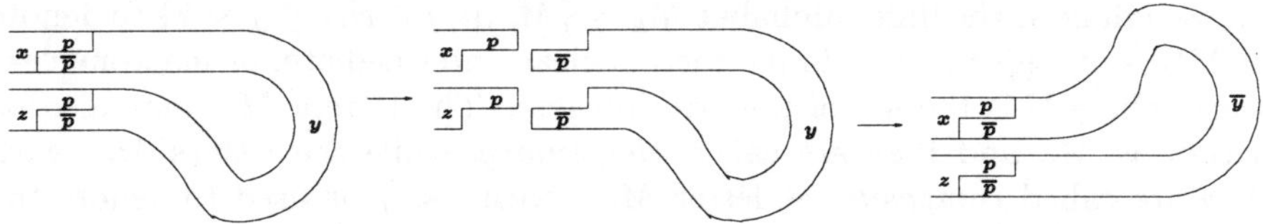

Figure 9. The *hi*-excision/reinsertion operation.

only one molecule. A bigger composite MDS is formed (as well as a bigger composite IES).

3. The operation of *dlad*-excision/reinsertion is applicable to molecules that have an *alternating direct repeat pattern* of the pair of pointers (p, q), i.e., molecules that have a direct repeat pattern (p, p) *and* a direct repeat pattern (q, q), and moreover one occurrence of q is between the two occurrences of p (and the first occurrence of p comes before the first occurrence of q). Such a molecule is folded into a double loop in such a way that the two occurrences of p are aligned within one loop, and the two occurrences of q are aligned within the other loop. Then, the operation proceeds as shown in Figure 10. Once again, the excision here involves staggered cuts, but through reinsertion, the operation yields one molecule only.

The molecular operations discussed in this section are from [13] and [36].

5 MDS structures and descriptors.

In the first formalization of a (micronuclear, intermediate, or macronuclear) gene that we will consider, we ignore the specific sequence of nucleotides

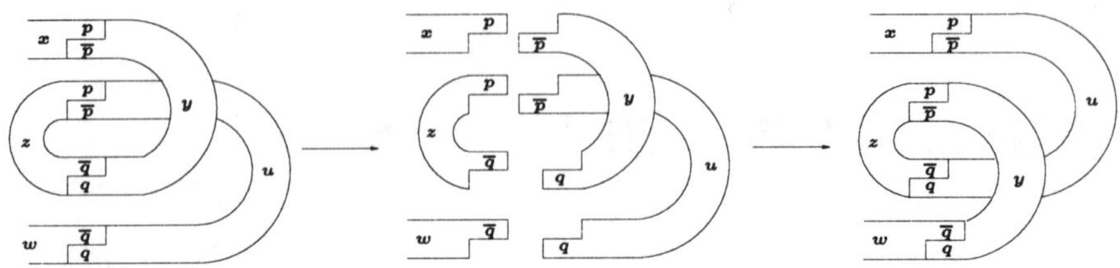

Figure 10. The *dlad*-excision/reinsertion operation.

forming a gene, and denote a gene simply by its sequence of MDSs. To this aim we will need the finite alphabet $\mathcal{M}_k = \{M_{i,j} | i, j \in \mathbb{N}, i \leq j \leq k\}$ to denote the MDSs of a given gene (in its micronuclear, intermediate, or macronuclear form), where $\mathbb{N}$ is the set of positive integers. The letters $M_{i,i}$ may also be written as M_i, and they are called *elementary*, while the letters $M_{i,j}$ with $i < j$ are called *composite*. A letter $M_{i,j}$, with $i < j$, is used to denote the composite MDS formed in the process of assembly by splicing the elementary MDSs $M_i, M_{i+1}, ..., M_j$ through their pointers. Thus, the incoming pointer of $M_{i,j}$ is i (or the marker b, if $i = 1$), and its outgoing pointer is $j + 1$ (or the marker e, if $j = \mathrm{k}$). We use the alphabet $\bar{\mathcal{M}}_k = \{\overline{M}_{i,j} | M_{i,j} \in \mathcal{M}_k\}$ to denote the inversions of the MDSs. Let $\Theta_k = \mathcal{M}_k \cup \bar{\mathcal{M}}_k$.

A sequence $M_{i_1,j_1}, M_{i_2,j_2}, ..., M_{i_n,j_n}$ is called *orthodox* if $i_i = 1, j_n = k$, and $i_l = 1 + j_{l-1}$, for all $2 \leq l \leq n$. A sequence over Θ_k is called a *real MDS structure* if it is obtained by permuting an orthodox sequence, possibly inverting some of its elements.

Using the alphabet Θ_k we can write down the MDS structure of the micronuclear, intermediate or the macronuclear gene in the form of a real MDS structure. Thus, e.g., the real MDS structure $M_1 M_6 \overline{M}_{2,5} M_7$ denotes the MDS structure of an intermediate gene such that the elementary MDSs M_2, M_3, M_4, M_5 have already been spliced together forming the composite MDS $M_{2,5}$, which is inverted in this intermediate gene. $M_1 M_2 M_3 M_4 M_5 M_6$ is the MDS structure of the micronuclear version of the $R1$ gene in *Oxytricha nova*, and $M_3 M_4 M_6 M_5 M_7 M_9 \overline{M}_2 M_1 M_8$ is the MDS structure of the micronuclear version of the actin I gene in *Oxytricha nova*, see [35].

The main idea behind our modeling of the assembly process is to keep track of pointers (and markers) *only*. Indeed, the pointers are used to align the fold required by each operation, and the assembly process continues as long as there are pointers left in the molecule, that is, as long as there are IESs left in the molecule (still separating MDSs). On the other hand, the macronuclear gene does not have IESs, and hence, it does not contain pointers. Hence, *when the pointers are eliminated, the assembly process is completed.*

In order to keep track of pointers only, we will denote each MDS $M = (p, \mu, q)$ by the ordered pair of its pointers/markers (p, q), obtaining in this way MDS descriptors. More formally, we define the mapping ψ on Θ_k by:

(i) $\psi(M_{1,k}) = (b, e)$, and $\psi(\overline{M}_{1,k}) = (\overline{e}, \overline{b})$,

(ii) $\psi(M_{1,i}) = (b, i+1)$, and $\psi(\overline{M}_{1,i}) = (\overline{i+1}, \overline{b})$, for all $1 \leq i < k$,

(iii) $\psi(M_{i,k}) = (i, e)$, and $\psi(\overline{M}_{i,k}) = (\overline{e}, \overline{i})$, for all $1 < i \leq k$,

(iv) $\psi(M_{i,j}) = (i, j+1)$, and $\psi(\overline{M}_{i,j}) = (\overline{j+1}, \overline{i})$, for all $1 < i \leq j < k$,

where b, e are reserved symbols. Then, for a sequence $X_1 \dots X_l$ over Θ_k,

$$\psi(X_1 \dots X_l) = \psi(X_1) \dots \psi(X_l).$$

Recall that the alphabet $\Pi_k = \{2, 3, \dots, k, \overline{2}, \overline{3}, \dots, \overline{k}\}$ denotes the pointers and their inversions. Also, let $\Psi = \{b, e, \overline{b}, \overline{e}\}$ be the set of markers, and let $\Pi_{ex,k} = \Pi_k \cup \Psi$. Then let Γ_k be the set of ordered pairs over $\Pi_{ex,k}$ consisting of:

- (b, e), $(\overline{e}, \overline{b})$,
- (b, i), $(\overline{i}, \overline{b})$, for all $2 \leq i \leq k$,
- (i, e), $(\overline{e}, \overline{i})$, for all $2 \leq i \leq k$,
- (i, j), $(\overline{j}, \overline{i})$, for all $2 \leq i < j \leq k$.

A string over Γ_k is called an MDS *descriptor*. Then, an MDS descriptor δ is *realistic*, if $\delta = \psi(x)$, for some real MDS structure x (over Θ_k).

Since each of our three molecular operations has a specific fold that is aligned by pointers, the information about pointers contained in realistic MDS descriptors is certainly sufficient for formalizing these operations in the framework of realistic MDS descriptors. One obtains in this way three rules **ld**, **hi**, and **dlad** defined on realistic MDS descriptors (corresponding to the molecular operations *ld*, *hi*, and *dlad* respectively). The definition of these rules is rather straightforward although somewhat tedious, because one has to take into account quite a number of different cases: for each pair of occurrences of pointers, such as (p, p) or $(p, \bar{p})$, each single occurrence is either an input or an output pointer, and, if one occurrence is an input (output) pointer, then

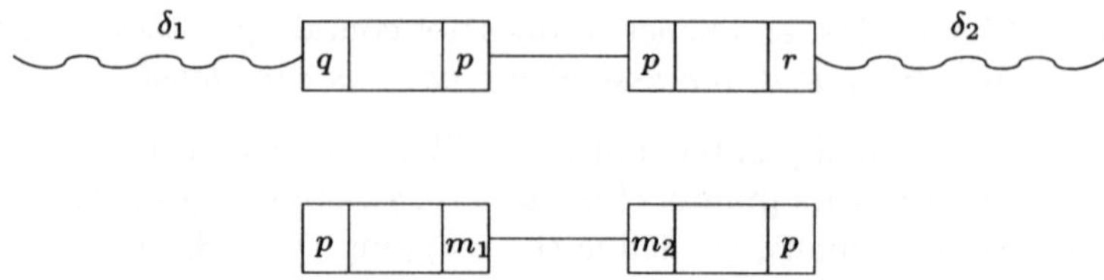

Figure 11. The two cases from the definition of $\mathbf{ld}_p$.

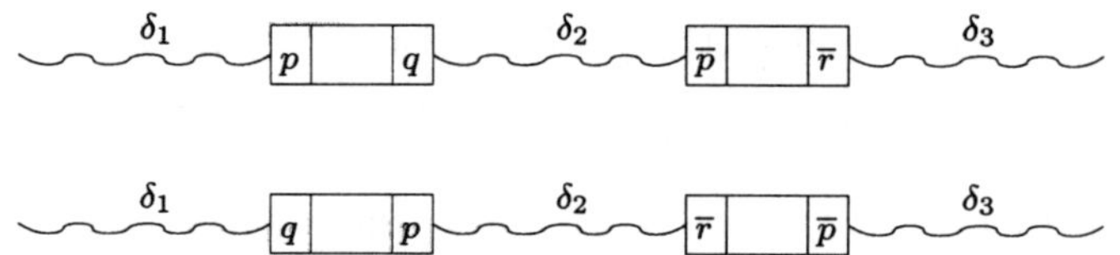

Figure 12. The two cases from the definition of $\mathbf{hi}_p$.

the other occurrence must be an output (input) pointer (the reader is referred to [6] for more intuition behind these rules).

1. For each $p \in \Pi_k$, the **ld**-rule for p, denoted by $\mathbf{ld}_p$, is defined as follows:

$$\begin{aligned}\mathbf{ld}_p(\delta_1(q,p)(p,r)\delta_2) &= \delta_1(q,r)\delta_2,\\ \mathbf{ld}_p((p,m_1)(m_2,p)) &= (m_2,m_1),\end{aligned}$$

where $q,r \in \Pi_{ex,k}$, $\delta_1,\delta_2 \in (\Gamma_k)^*$, and $m_1,m_2 \in \Psi$. The two cases are illustrated in Figure 11.

2. For each $p \in \Pi_k$, the **hi**-rule for p, denoted by $\mathbf{hi}_p$, is defined as follows:

$$\begin{aligned}\mathbf{hi}_p(\delta_1(p,q)\delta_2(\overline{p},\overline{r})\delta_3) &= \delta_1\,\mathbf{rs}(\delta_2)(\overline{q},\overline{r})\delta_3,\\ \mathbf{hi}_p(\delta_1(q,p)\delta_2(\overline{r},\overline{p})\delta_3) &= \delta_1(q,r)\,\mathbf{rs}(\delta_2)\delta_3,\end{aligned}$$

where $q,r \in \Pi_{ex,k}$ and $\delta_1,\delta_2 \in (\Gamma_k)^*$, and **rs** is the operation on $(\Gamma_k)^*$ defined by: $\mathbf{rs}((p_1,q_1)...(p_l,q_l)) = (\bar{q}_l,\bar{p}_l)...(\bar{q}_1,\bar{p}_1)$ for any $l \geq 0$. The two cases are illustrated in Figure 12.

3. For each $p,q \in \Pi_k$, $p \neq q$, the **dlad** rule for p and q, denoted as $\mathbf{dlad}_{p,q}$, is

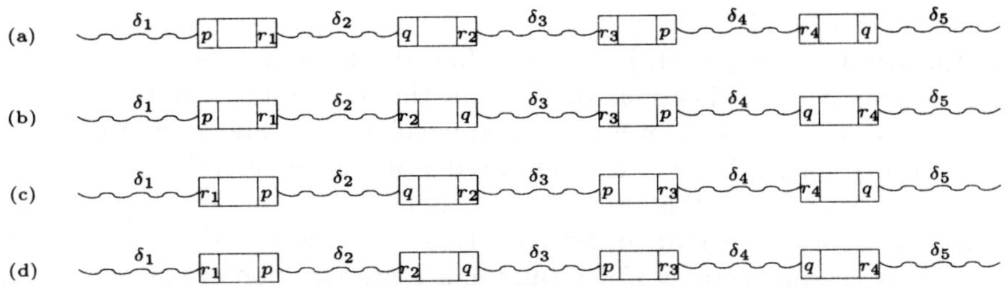

Figure 13. The four main cases from the definition of $\mathbf{dlad}_{p,q}$.

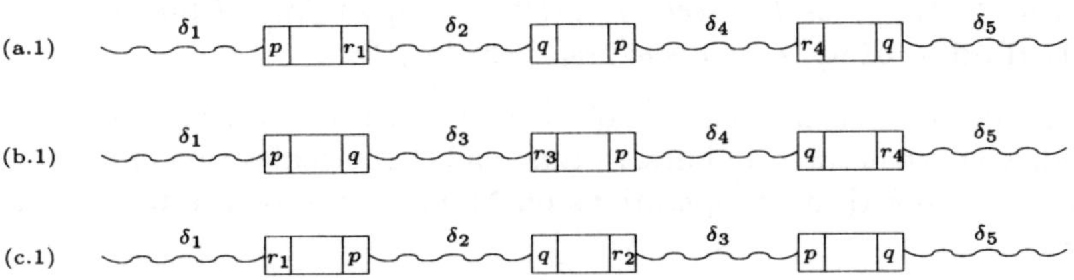

Figure 14. The three special cases from the definition of $\mathbf{dlad}_{p,q}$.

defined as follows:

(a) $\mathbf{dlad}_{p,q}(\delta_1(p,r_1)\delta_2(q,r_2)\delta_3(r_3,p)\delta_4(r_4,q)\delta_5) = \delta_1\delta_4(r_4,r_2)\delta_3(r_3,r_1)\delta_2\delta_5$,
$(a.1)$ $\mathbf{dlad}_{p,q}(\delta_1(p,r_1)\delta_2(q,p)\delta_4(r_4,q)\delta_5) = \delta_1\delta_4(r_4,r_1)\delta_2\delta_5$,
(b) $\mathbf{dlad}_{p,q}(\delta_1(p,r_1)\delta_2(r_2,q)\delta_3(r_3,p)\delta_4(q,r_4)\delta_5) = \delta_1\delta_4\delta_3(r_3,r_1)\delta_2(r_2,r_4)\delta_5$,
$(b.1)$ $\mathbf{dlad}_{p,q}(\delta_1(p,q)\delta_3(r_3,p)\delta_4(q,r_4)\delta_5) = \delta_1\delta_4\delta_3(r_3,r_4)\delta_5$,
(c) $\mathbf{dlad}_{p,q}(\delta_1(r_1,p)\delta_2(q,r_2)\delta_3(p,r_3)\delta_4(r_4,q)\delta_5) = \delta_1(r_1,r_3)\delta_4(r_4,r_2)\delta_3\delta_2\delta_5$,
$(c.1)$ $\mathbf{dlad}_{p,q}(\delta_1(r_1,p)\delta_2(q,r_2)\delta_3(p,q)\delta_5) = \delta_1(r_1,r_2)\delta_3\delta_2\delta_5$,
(d) $\mathbf{dlad}_{p,q}(\delta_1(r_1,p)\delta_2(r_2,q)\delta_3(p,r_3)\delta_4(q,r_4)\delta_5) = \delta_1(r_1,r_3)\delta_4\delta_3\delta_2(r_2,r_4)\delta_5$,

where $r_1, r_2, r_3, r_4, r_5 \in \Pi_{ex,k}$, and $\delta_1, \delta_2, \delta_3, \delta_4, \delta_5 \in (\Gamma_k)^*$. In each of the above instances of $\mathbf{dlad}_{p,q}$, the pointer p overlaps with the pointer q, i.e., the molecule represented by the realistic MDS descriptor to which $\mathbf{dlad}_{p,q}$ is applied, has a required double repeat pattern (p,q,p,q). The seven cases are illustrated in Figures 13 and 14.

Example 1 *For the realistic MDS descriptors* $\delta_1 = (2,3)(3,4)(6,e)(5,6)(b,2)(4,5)$ *and* $\delta_2 = (4,5)(\bar{3},\bar{2})(b,2)(3,4)(5,e)$, *we have*
$\mathbf{ld}_3(\delta_1) = (2,4)(6,e)(5,6)(b,2)(4,5)$,
$\mathbf{dlad}_{2,5}(\delta_1) = (4,6)(b,3)(3,4)(6,e)$,
$\mathbf{hi}_{\bar{3}}(\delta_2) = (4,5)(\bar{2},\bar{b})(2,4)(5,e)$.

Now for a realistic MDS descriptor δ and rules $\rho_1, ..., \rho_l$, $l \geq 1$, from the set $\{\mathbf{ld}_p, \mathbf{hi}_p, \mathbf{dlad}_{p,q} : p, q \in \Pi_k\}$ we can define the reduction of δ by $\rho_1, ..., \rho_l$ as the sequence of applications of $\rho_1, ..., \rho_l$ (in this order). More formally, $(\delta; \rho_1, ..., \rho_l)$ is a *reduction* of δ if ρ_1 is applicable to δ, and ρ_i is applicable to $\rho_{i-1}..\rho_1(\delta)$, for all $1 < i \leq l$. Such a reduction is *successful* if $\rho_l...\rho_1(\delta)$ is either (b, e), or $(\bar{e}, \bar{b})$. In this case, we also say that $(\delta; \rho_1, ..., \rho_l)$ is a *successful strategy* for δ (*based on* { **ld,hi,dlad** }). Hence a sequence of applications of rules is a successful strategy if it results in a representation of one composite MDS beginning with b and ending with e (such a representation may be either (b, e) or $(\bar{e}, \bar{b})$). Clearly, this corresponds to a successful gene assembly.

Example 2 *Let δ_1 be the realistic MDS descriptor from Example 1. Then $(\delta_1; \mathbf{ld}_3, \mathbf{dlad}_{6,5}, \mathbf{dlad}_{2,4})$ is a successful strategy for δ_1.*

The formalisation of the structure of the micronuclear and intermediate genes through MDS structures and descriptors, and the formalisation of the process of gene assembly through operations on MDS descriptors, is from [13], [11] and [8].

6 String pointer reduction systems.

Realistic MDS descriptors code quite precisely the IES/MDS structure of micronuclear, intermediate and macronuclear genes. This is achieved by the use of the alphabet Π_k of pointers, the alphabet Ψ of markers, and the parentheses "(" and ")". The basic intuition is that a realistic MDS descriptor is an "instantaneous description" (a snapshot) of the IES/MDS structure of a gene at some point in the process of gene assembly. In this section we make another abstraction step by elimination of markers and parentheses, retaining only pointers for writing instantaneous descriptions.

For a realistic MDS descriptor δ by erasing markers, parentheses "(" and ")" (and commas!) we obtain the *legal string π_δ corresponding to δ*. We use ϕ to denote this translation, i.e., $\phi(\delta) = \pi_\delta$. Thus, for the realistic MDS descriptor $\delta = (\bar{3}, \bar{2})(4, 5)(b, 2)(\bar{4}, \bar{3})(6, e)(5, 6)$, we get $\phi(\delta) = \bar{3}\bar{2}452\bar{4}\bar{3}656$. The legal string π_δ corresponding to δ, is *more abstract* than δ – it carries less information. In more formal terms, ϕ is *not* injective. For example, two different realistic MDS descriptors
$\delta_1 = (b, 2)(3, 4)(5, 6)(2, 3)(4, 5)(6, e)$, and
$\delta_2 = (2, 3)(4, 5)(6, e)(b, 2)(3, 4)(5, 6)$,
yield the same legal string 2345623456.

Thus, by translating realistic MDS descriptors we obtain legal strings over

Π_k, which are defined as follows. A string $\pi \in \Pi_k^*$ is called *legal* if for each $p \in \Pi_k$, if π contains an occurrence of a pointer from $\mathbf{pts}(p)$, then π contains exactly two occurrences of pointers from $\mathbf{pts}(p)$; thus either π contains one occurrence of p and one occurrence of $\bar{p}$ (and we say that $\mathbf{pts}(p)$ is *positive* in π), or π contains two occurrences of p or π contains two occurrences of $\bar{p}$ (and we say that $\mathbf{pts}(p)$ is *negative* in π). If p occurs in π, then we say that p is *positive* (*negative*) in π if $\mathbf{pts}(p)$ is positive (negative, resp.) in π. The set of legal strings over Π_k is bigger than the set of legal strings obtained from realistic MDS descriptors by translating them by ϕ - the so obtained legal strings are called *realistic*. Thus, e.g., the legal string 25352434 is *not* realistic, while the legal string $25\bar{5}\bar{3}23$ is realistic: it is obtained by ϕ from the realistic MDS descriptor $(b,2)(5,e)(\bar{5},\bar{3})(2,3)$.

We will define now the string pointer reduction system over Π_k - it formalizes the effect of molecular operations *ld*, *hi* and *dlad*, but now on the level of legal strings. First, we introduce a couple of auxiliary notions.

Let $\pi = x_1 ... x_n \in \Pi_k^*$ with $x_1, ..., x_n \in \Pi_k$. For a pointer $p \in \Pi_k$ such that $\{x_i, x_j\} \subseteq pts(p)$ for some positive integers $i < j \leq n$, the *p-interval of* π is the substring $x_i x_{i+1} ... x_j$. Pointers $p, q \in \Pi_k$ *overlap in* π if the p-interval of π overlaps with the q-interval of π (to recall the notion of overlapping substrings, see Section 1) – we also say then that $\mathbf{pts}(p)$ overlaps with $\mathbf{pts}(q)$. For example, for the legal string $\pi = \bar{3}245 2\bar{4}3656$ (obtained above by applying ϕ), the 2-interval is $\bar{2}452$, the 4-interval is $452\bar{4}$, and the 6-interval is 656. Hence pointers 2 and 4 overlap in π, but pointers 2 and 6 do not (and neither do pointers 4 and 6). We define the *reversed switch of* π, denoted $\mathbf{rs}(\pi)$, to be the string $\bar{x}_n ... \bar{x}_1$. For example, $\mathbf{rs}(\bar{2}452\bar{4}) = 4\bar{2}\bar{5}\bar{4}2$.

The string pointer reduction system over Π_k consists of a number of rules for rewriting legal strings – these rules are defined as follows.

Let π be a legal string over Π_k and let $p, q \in \Pi_k$.

The *string negative rule* for p, denoted by $\mathbf{snr}_p$, is applicable to π if pp is a substring of π, i.e., $\pi = \pi_1 pp \pi_2$, for some strings π_1, π_2 over Π_k. The result of the application of $\mathbf{snr}_p$ to π is the legal string

$$\mathbf{snr}_p(\pi) = \pi_1 \pi_2.$$

The *string positive rule* for p, denoted by $\mathbf{spr}_p$, is applicable to π if $p\bar{p}$ is a scattered substring of π, i.e., $\pi = \pi_1 p \pi_2 \bar{p} \pi_3$, for some strings π_1, π_2, π_3 over Π_k. The result of the application of $\mathbf{spr}_p$ to π is the legal string

$$\mathbf{spr}_p(\pi) = \pi_1 \, \mathbf{rs}(\pi_2) \pi_3.$$

The *string double rule* for p and q, denoted by $\mathbf{sdr}_{p,q}$, is applicable to π if the pointers p and q overlap in π, with the first occurrence of p preceding the first occurrence of q, i.e., $\pi = \pi_1 p \pi_2 q \pi_3 p \pi_4 q \pi_5$, for some strings $\pi_1, \pi_2, \pi_3, \pi_4, \pi_5$ over Π_k. The result of the application of $\mathbf{sdr}_{p,q}$ to π is the legal string

$$\mathbf{sdr}_{p,q}(\pi) = \pi_1 \pi_4 \pi_3 \pi_2 \pi_5.$$

Example 3 *Consider the legal string* $\pi = \bar{7}\bar{6}3423246788$. *Then, the rules* $\mathbf{snr}_8, \mathbf{spr}_6$, *and* $\mathbf{sdr}_{3,4}$ *are applicable to* π, *and the results of applying these rules to* π *are:* $\mathbf{snr}_8(\pi) = \bar{7}\bar{6}34232467$, $\mathbf{spr}_6(\pi) = \bar{7}\bar{4}\bar{2}\bar{3}\bar{2}\bar{4}\bar{3}788$, *and* $\mathbf{sdr}_{3,4}(\pi) = \bar{7}\bar{6}226788$.

The *string pointer reduction system over* Π_k, denoted by $\mathbf{SPRS}_{\Pi_k}$, or simply $\mathbf{SPRS}$ if Π_k is understood from the context of considerations, is the set of the above operations. Thus, $\mathbf{SPRS} = \{\mathbf{snr}_p, \mathbf{spr}_p, \mathbf{sdr}_{p,q} \mid p, q \in \Pi_k\}$.

Again, in order to formalize the process of gene assembly through a sequence of applications of the molecular operations *ld*, *hi* and *dlad*, we define a reduction of a legal string by a sequence of string pointer reduction rules as follows. For a legal string π and a sequence of rules $\rho_1, \ldots, \rho_l$, $l \geq 1$, from $\mathbf{SPRS}$, $D = (\pi; \rho_1, \ldots, \rho_l)$ is a *reduction (of* π *by* $\rho_1, \ldots, \rho_l$*)*, if ρ_1 is applicable to π, and ρ_i is applicable to $\rho_{i-1} \ldots \rho_1(\pi)$, for all $1 < i \leq l$. The *result of* D is the legal string $\rho_l \ldots \rho_1(\pi)$. We say that D is *successful*, if the result of D is the empty string.

Example 4 *For the legal string* π *from Example 3,* $D = (\pi; \mathbf{spr}_6, \mathbf{snr}_8, \mathbf{spr}_7, \mathbf{sdr}_{3,4}, \mathbf{snr}_2)$ *is a successful reduction, and the corresponding sequence of strings starting with* π *and ending with* Λ *is:*
$\bar{7}\bar{6}3423246788$,
$\bar{7}\bar{4}\bar{2}\bar{3}\bar{2}\bar{4}\bar{3}788$,
$\bar{7}\bar{4}\bar{2}\bar{3}\bar{2}\bar{4}\bar{3}7$
342324,
22,
Λ.

Note that for a legal string π, if either one of the rules $\mathbf{snr}_p$ or $\mathbf{spr}_p$ is applicable to π, for some $p \in \Pi_k$, then both occurrences in π from $\mathbf{pts}(p)$ are removed as the result of the reduction of π by that rule. Also, if $\mathbf{sdr}_{p,q}$ is applicable to π, for some $p, q \in \Pi_k$, then all occurrences in π from $\mathbf{pts}(p)$ and from $\mathbf{pts}(q)$ are removed as the result of the reduction of π by $\mathbf{sdr}_{p,q}$.

In the previous section we formalized the molecular operations *ld*, *hi* and *dlad* through the formal operations **ld**, **hi** and **dlad**, respectively, which operate in the framework of MDS descriptors. In this section we have formalized the same molecular operations *ld*, *hi* and *dlad* through the formal operations **spr**, **snr**, and **sdr**, respectively, which operate in a more abstract framework of legal words. However, it turns out that the two frameworks are translatable into each other in the sense of the following result.

Theorem 1 *Let δ be a realistic MDS descriptor.*

(i) Let $p \in \Pi_k$. The rule $\mathbf{ld}_p$ is applicable to δ if and only if $\mathbf{snr}_p$ is applicable to π_δ. Moreover, in the case of applicability, if $\pi' = \mathbf{snr}_p(\pi_\delta)$ and $\delta' = \mathbf{ld}_p(\delta)$, then $\pi' = \pi_{\delta'}$.

(ii) Let $p \in \Pi_k$. The rule $\mathbf{hi}_p$ is applicable to δ if and only if $\mathbf{spr}_p$ is applicable to π_δ. Moreover, in the case of applicability, if $\pi' = \mathbf{spr}_p(\pi_\delta)$ and $\delta' = \mathbf{hi}_p(\delta)$, then $\pi' = \pi_{\delta'}$.

(iii) Let $p, q \in \Pi_k$ be two distinct pointers. The rule $\mathbf{dlad}_{p,q}$ is applicable to δ if and only if $\mathbf{sdr}_{p,q}$ is applicable to π_δ. Moreover, in the case of applicability, if $\pi' = \mathbf{sdr}_{p,q}(\pi_\delta)$ and $\delta' = \mathbf{dlad}_{p,q}(\delta)$, then $\pi' = \pi_{\delta'}$.

Thus, the successful strategies for realistic MDS descriptors, and the successful reductions of legal strings are intertranslatable.

This section is based on [13], [11], and [8].

7 Graph pointer reduction systems.

The main idea of our approach to formalize the gene assembly process is to keep track of pointers – our snapshots (instantaneous descriptions) give the pattern of pointers present in a given micronuclear or intermediate gene. The legal string π_δ corresponding to a realistic MDS descriptor δ gives exactly the sequence of pointers in (the IES/MDS structure described by) δ. In this section we will consider less precise snapshots: we will record the pointer sets present at a given moment (and for each pointer set the information of whether it is positive or negative) as well as the overlapping relationship between these pointer sets. This is achieved by considering signed overlap graphs which form our next abstraction level in formalizing the gene assembly process.

Let π be a legal string over Π_k. The *signed overlap graph of* π is the signed graph $\gamma_\pi = (V, E, \theta)$ such that:
$V = \{\mathbf{pts}(p) : p \in \Pi_k \text{ and } p \text{ occurs in } \pi\}$,
$E = \{\{v, u\} : v \neq u, v = \mathbf{pts}(p), u = \mathbf{pts}(q)$ for some $p, q \in \Pi_k$ that overlap in $\pi\}$,
for each $v \in V, \theta(v) = +$ iff v is positive in π.

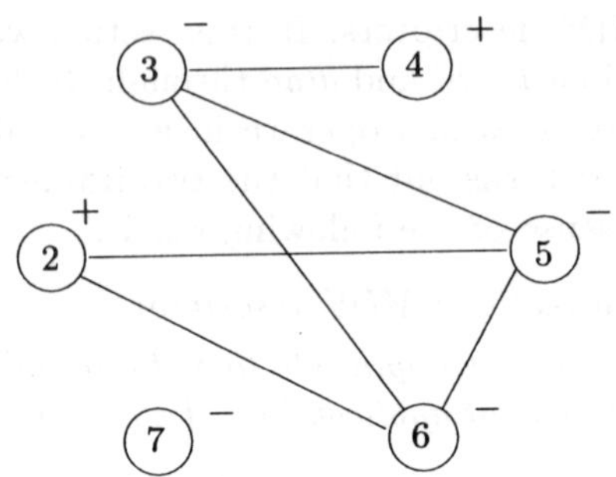

Figure 15. The signed overlap graph γ_π of π.

Example 5 *Consider the legal string* $\pi = \bar{2}\bar{4}3456327756$. *Then the signed overlap graph* γ_π *of* π *is given in Figure 15.*

In order to simplify Figure 15, we indicate the identity of each node by writing p rather than $\mathbf{pts}(p)$. As a matter of fact, we will use this simplified notation in the sequel of this section by writing p also when we refer to the node $\mathbf{pts}(p)$ – this identification will always be obvious from the context of considerations.

Again, the signed overlap graph γ_π of π is *more abstract* than π – it carries less information: many different legal strings may yield the same signed overlap graph. For example, the legal strings $42233\bar{4}, 2334\bar{4}2, 24\bar{4}332, 3224\bar{4}3$ (and a number of others) yield the same (very simple!) signed overlap graph. Thus, in going from realistic MDS descriptors to legal strings, and then to signed overlap graphs we increase the abstract level of our considerations – we get rid of more and more "technical details".

The graph pointer reduction system over Π_k consists of a number of rules for rewriting signed overlap graphs – these rules are defined as follows.

Let π be a legal string over Π_k, γ_π its signed overlap graph, and $p, q \in \Pi_k$.

The *graph negative rule* for p, denoted by $\mathbf{gnr}_p$, is applicable to γ_π if the node p is isolated and negative in γ_π. The result of the application of $\mathbf{gnr}_p$ to γ_π is the signed overlap graph $\mathbf{gnr}_p(\gamma_\pi)$, obtained from γ_π by removing the node p.

The *graph positive rule* for p, denoted by $\mathbf{gpr}_p$, is applicable to γ_π if the node p is positive in γ_π. The result of the application of $\mathbf{gpr}_p$ to γ_π is the signed overlap graph $\mathbf{gpr}_p(\gamma_\pi)$, obtained from $\mathbf{lcom}_x(\gamma_\pi)$ by removing the node p

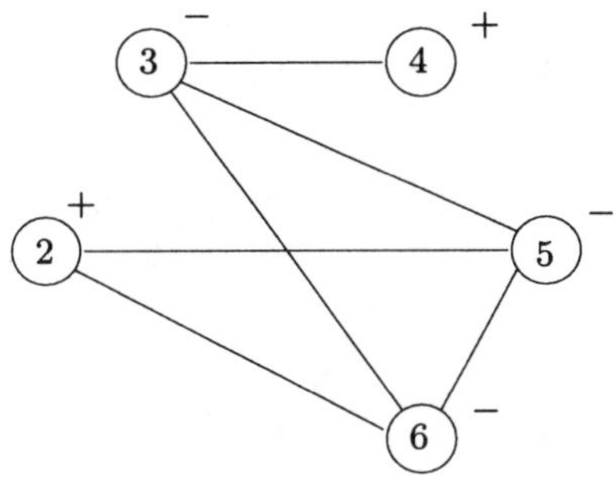

Figure 16. The sign graph $\mathbf{gnr}_7(\gamma_\pi)$.

(recall that $\mathbf{lcom}_x(\gamma_\pi)$ is the local complementation of γ_π at p – see Section 1).

The *graph double rule* for p and q, denoted by $\mathbf{gdr}_{p,q}$, is applicable to γ_π if the nodes p and q are negative and adjacent in γ_π. If $\gamma_\pi = (V, E, \theta)$, then the result of the application of $\mathbf{gdr}_{p,q}$ to γ_π is the signed overlap graph $\mathbf{gdr}_{p,q}(\gamma_\pi) = (V \setminus \{p,q\}, E', \theta')$ obtained as follows: θ' equals θ restricted to $V \setminus \{p,q\}$, and E' is obtained from E by complementing the edges *between* the sets $\mathbf{ngb}_{\gamma_\pi}(p)$ and $\mathbf{ngb}_{\gamma_\pi}(q)$. This means that the status of a pair $\{x,y\}$ (for $x, y \in V \setminus \{p,q\}$) as an edge will change if and only if $x \in \mathbf{ngb}_{\gamma_\pi}(p), y \in \mathbf{ngb}_{\gamma_\pi}(q)$ and *not* both x and y are in $\mathbf{ngb}_{\gamma_\pi}(p) \cap \mathbf{ngb}_{\gamma_\pi}(q)$.

Example 6 *For the signed graph γ_π from Example 5, see Figure 15, the signed graphs $\mathbf{gnr}_7(\gamma_\pi), \mathbf{gpr}_2(\gamma_\pi)$ and $\mathbf{gdr}_{3,6}(\gamma_\pi)$ are given in Figures 16, 17 and 18, respectively.*

Now, the *graph pointer reduction system* over Π_k, denoted by $\mathbf{GPRS}_{\Pi_k}$, or simply **GPRS** if Π_k is understood from the context of considerations, is the set of the above reduction rules. Thus, $\mathbf{GPRS} = \{\mathbf{gnr}_p, \mathbf{gpr}_p, \mathbf{gdr}_{p,q} \mid p, q \in \Pi_k\}$.

Again, we formalize the process of gene assembly by defining a reduction of a signed overlap graph by a sequence of graph pointer reduction rules as follows.

Let π be a legal string and γ_π its signed overlap graph. For a sequence of reduction rules $\rho_1, \ldots, \rho_l$, $l \geq 1$, from **GPRS**, $D = (\gamma_\pi; \rho_1, \ldots, \rho_l)$ is a *reduction* (of γ_π by $\rho_1, \ldots, \rho_l$), if ρ_1 is applicable to γ_π, and ρ_i is applicable to $\rho_{i-1} \ldots \rho_1(\gamma_\pi)$, for all $1 < i \leq l$. The *result of D* is the signed overlap graph $\rho_l \ldots \rho_1(\gamma_\pi)$. We say that D is *successful* if the result of D is the empty graph.

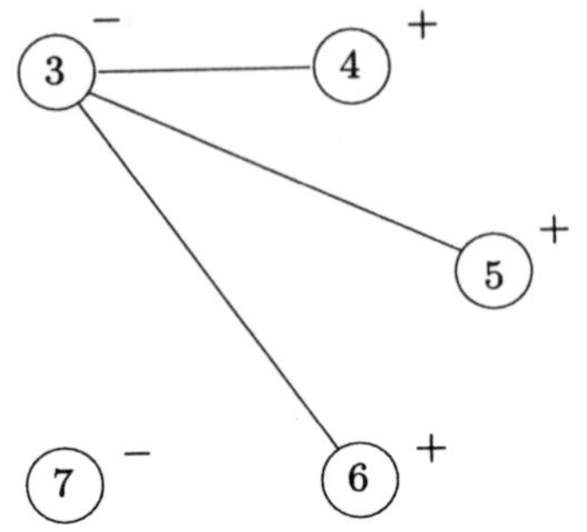

Figure 17. The sign graph $\mathbf{gpr}_2(\gamma_\pi)$.

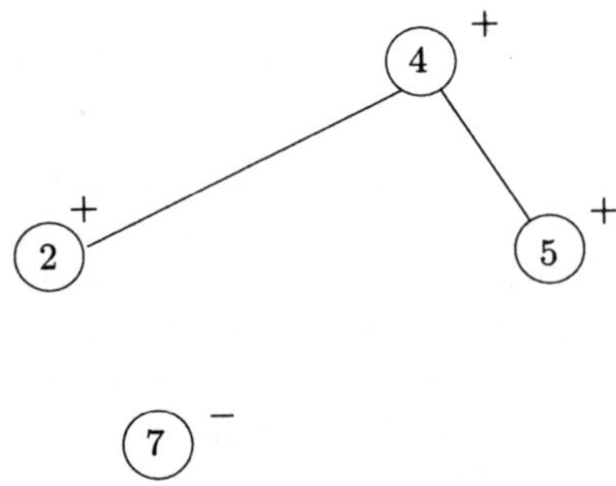

Figure 18. The sign graph $\mathbf{gdr}_{3,6}(\gamma_\pi)$.

Example 7 *The signed overlap graph* γ_π *from Example 5, see Figure 15, is successful, and* $(\gamma_\pi; \mathbf{gdr}_{3,6}, \mathbf{gnr}_7, \mathbf{gpr}_2, \mathbf{gpr}_5, \mathbf{gpr}_4)$ *is a successful reduction.*

In this section we have formalized the molecular operations ld, hi, and $dlad$, through graph pointer reduction rules **gnr**, **gpr** and **gdr**, respectively, operating on signed overlap graphs. It turns out that this framework and the framework of string pointer reduction systems are mutually translatable – with the rules **snr**, **spr**, and **sdr** intertranslatable with the rules **gnr**, **gpr**, and **gdr**, respectively. This is expressed by the following two results.

Theorem 2 *Every string reduction* $D = (\pi; \rho_1, \ldots, \rho_l)$ *in* **SPRS** *translates into a graph reduction* $D' = (\gamma_\pi; \rho'_1, \ldots, \rho'_l)$ *in* **GPRS**, *by translating:*

- *each* **SPRS** *reduction rule* $\mathbf{snr}_p$ *into the* **GPRS** *reduction rule* $\mathbf{gnr}_p$*;*
- *each* **SPRS** *reduction rule* $\mathbf{spr}_p$ *into the* **GPRS** *reduction rule* $\mathbf{gpr}_p$*;*

- *each* **SPRS** *reduction rule* $\mathbf{sdr}_{p,q}$ *into the* **GPRS** *reduction rule* $\mathbf{gdr}_{p,q}$.

Consequently, if D is successful, then so is D'.

Theorem 3 *Let π be a legal string. Then, for every successful graph reduction $D' = (\gamma_\pi; \rho_1, \ldots, \rho_l)$ of γ_π in* **GPRS**, *there exists a permutation $D = (\gamma_\pi; \rho_{i_1}, \ldots, \rho_{i_l})$ of D' that can be translated into a successful string reduction $(\pi; \rho'_{i_1}, \ldots, \rho'_{i_l})$ in* **SPRS**, *by translating*

- *each* **GPRS** *reduction rule* $\mathbf{gnr}_p$ *into the* **SPRS** *reduction rule* $\mathbf{snr}_p$;
- *each* **GPRS** *reduction rule* $\mathbf{gpr}_p$ *into the* **SPRS** *reduction rule* $\mathbf{spr}_p$;
- *each* **GPRS** *reduction rule* $\mathbf{gdr}_{p,q}$ *into the* **SPRS** *reduction rule* $\mathbf{sdr}_{p,q}$.

While the translation from **SPRS** into **GPRS** (Theorem 2) is rather straightforward, the translation from **GPRS** into **SPRS** (Theorem 3) is more subtle. The problem is that in translating a legal string π into its sign overlap graph γ_π we ignore the exact order of pointers in π. Hence, it may happen that a pointer p in γ_π can be reduced, while it is not ready yet for reduction in π. For example, the legal string $\pi = pqqp$ translates into the signed overlap graph γ_π consisting of two negative isolated nodes, one corresponding to p and the other to q. One can reduce these two nodes in any order, e.g., first p and then q. However this order of reduction is impossible for π in **SPRS**: one has to apply $\mathbf{snr}_q$ first (qq is a substring of π) – this creates the (sub)string pp which can *then* be reduced by $\mathbf{snr}_p$. Therefore, one may need to permute the applications of some rules in **GPRS** before translating a successful reduction in **GPRS** into a successful reduction in **SPRS**.

This section is based on [13], [11], and [8].

8 Universality and simplicity.

The model of gene assembly in ciliates based on the three molecular operations: *ld, hi* and *dlad*, has turned out to be very successful in the sense that it can be applied to *all* experimental data on gene assembly in ciliates that are available *now*, see [36]. The question arises then whether a *future* experiment could discover a micronuclear MDS/IES pattern that cannot be assembled through our three molecular operations. The answer is NO, because one can prove that any micronuclear MDS/IES pattern can be assembled using these operations – in other words the set of three operations is *universal*. Formally, we have the following result.

Theorem 4 *Any realistic MDS descriptor has a successful strategy based on* $\{\mathbf{ld}, \mathbf{hi}, \mathbf{dlad}\}$.

This universality result is stated in terms of realistic MDS descriptors, but through our intertranslatability results it can be also stated in the form of legal strings or signed overlap graphs.

One needs all three operations for the universality. Hence, e.g., the realistic MDS descriptor $(b,2)(2,e)$ has no successful strategy in $\{\mathbf{hi}, \mathbf{dlad}\}$, the realistic MDS descriptor $(\bar{2},\bar{b})(2,e)$ has no successful strategy in $\{\mathbf{ld}, \mathbf{dlad}\}$, and the realistic MDS descriptor $(b,2)(3,e)(2,3)$ has no successful strategy in $\{\mathbf{ld}, \mathbf{hi}\}$.

When a molecular operation (*ld, hi,* or *dlad*) is applied to a molecule, then the parts of the molecule that are affected by the operation can be quite intricate. One way to measure the "complexity" of the affected parts is to take into account the number of IESs included in these parts, because in this way we measure the number of the already assembled (composite) MDSs involved in the operation. The most elementary case is when we restrict this number of IESs to one – in this way we get *simple ld*, *simple hi*, and *simple dlad* operations.

As a matter of fact, when we introduced the *ld* operation in Section 3, we restricted it right away to the simple case: there is exactly one IES between the two direct repeats of pointer p. This was absolutely necessary because if there were more than one IES between the two (nonboundary) direct repeats of pointer p, then at least one MDS would be excised (on a circular molecule) and a successful gene assembly could not be accomplished, because at least one MDS would be missing!

An application of *hi* to a hairpin-folded molecule aligned by an inverted repeat $(p,\bar{p})$ is *simple* if there is exactly one IES between p and $\bar{p}$. Such an application creates one new composed MDS from two existing MDSs which are adjacent (through an IES) with one of them inverted w.r.t. the other one. Thus, e.g., in the formalism of MDS descriptors, such an application converts an MDS descriptor $\alpha(r,p)(\bar{q},\bar{p})\beta$ for some pointers $r,\bar{q}$ into the MDS descriptor $\alpha(r,q)\beta$.

An application of *dlad* to a double folded molecule aligned by direct repeats (p,p) and (q,q) is *simple* if the segment of the molecule starting with the first occurrence of p and ending with the first occurrence of q is an MDS, and the segment of the molecule in between the second occurrence of p and the second occurrence of q is an IES. Recall that the net effect of an application of the *dlad* operation is the *exchange* of the part of the molecule in between the first occurrence of p and the first occurrence of q with the part of the molecule in between the second occurrence of p and the second occurrence of q. Therefore the simple application of *dlad* uses the MDS beginning with the first occurrence of p and ending with the first occurrence of q, to connect (to bridge together) the MDS ending on the second occurrence of p with the MDS beginning with the second occurrence of q. Thus in the formalism of MDS descriptors this application of simple *dlad* corresponds to the transformation of an MDS descriptor $\alpha(p,q)\beta(r,p)(q,s)\gamma$, for some pointers, r, q, into the MDS descriptor $\alpha\beta(r,s)\gamma$.

It turns out that we can explain all experimental data on gene assembly in ciliates available now using *only* simple operations, see [36]. However, one cannot guarantee *now* that the experimental data obtained in the *future* will also be explainable using only simple operations, because the set of simple versions of *ld, hi* and *dlad* is not universal. Formally we have the following result.

Theorem 5 *There exists realistic MDS descriptors which have no successful strategy based on the simple applications of* **ld**, **hi** *and* **dlad**.

This section is based on [10].

9 Patterns of micronuclear genes.

Understanding the MDS/IES structure of micronuclear genes is an important goal of research on gene assembly. By Theorem 4, the set of all three molecular operations *ld, hi,* and *dlad* can assemble the macronuclear version of each micronuclear gene. However, inspecting the MDS/IES structure of known micronuclear genes, one notices (see [35]) that some of these genes may be assembled using only a subset of the three operations. Thus, e.g., the βTP micronuclear gene in *Oxytricha trifallax* can be assembled using only the *ld* operation, while the αTP micronuclear gene in *Oxytricha trifallax* can be assembled using only *ld* and *dlad* operations. Also, the assembly process is intrinsically nondeterministic and in general there are many possible strategies to assemble a given micronuclear gene, where different strategies may use even different subsets of $\{ld, hi, dlad\}$. Thus, if, e.g., one wants to establish the "simplest" (most parsimonious) subset $\mathcal{S}$ of $\{ld, hi, dlad\}$ that can assemble a given gene τ, then one would like to know the *domain* of $\mathcal{S}$, i.e., the set of *all* MDS/IES patterns of micronuclear genes that can be assembled by $\mathcal{S}$ (and then check whether or nor τ belongs to this domain). As a matter of fact, for each subset $\mathcal{S}$ of $\{ld, hi, dlad\}$, there exists a full characterization of

its domain. We will give some of these results here – they are formulated in terms of realistic (legal) strings. The reader is referred to [7] for the whole landscape of domains.
We begin by looking at single operations (i.e., $\mathcal{S}$ is a singleton set). Recall that, unless stated otherwise, we assume that the micronuclear gene contains k MDSs; we will also assume that, unless stated otherwise, a realistic string is a translation (by π) of a real MDS structure of size k. We will use the following notation:
$LD_k = \{\mathbf{snr}_p : p \in \Pi_k\}$,
$HI_k = \{\mathbf{spr}_p : p \in \Pi_k\}$, and
$DLAD_k = \{\mathbf{sdr}_{p,q} : p, q \in \Pi_k\}$.

Theorem 6 *A realistic string u has a successful reduction in LD_k if and only if $u \in Z \cup \overline{Z}$, where*
$Z = \{2233...kk\} \cup \{p(p+1)(p+1)...kk22...(p-1)(p-1)p : 2 \leq p \leq k\}$, *and*
$\overline{Z} = \{\overline{w} : w \in Z\}$,
where for a string $w = x_1...x_n$, with $x_i \in \Pi_k$ for $1 \leq i \leq n, \overline{w} = \overline{x}_n...\overline{x}_2\overline{x}_1$.

We need some auxiliary notions for stating the other results.
Let $P = \{\alpha_1, ..., \alpha_n\}, n \geq 1$, be a set of strings over Π_k. We say that P is *legal* if $\alpha_{i_1}...\alpha_{i_n}$ is legal for some permutation $i_1, ..., i_n$ of $1, ..., n$. Also, $p \in \Pi_k$ *occurs in* P (is *negative* in P, is *positive* in P) if p occurs in $\alpha_{i_1}...\alpha_{i_n}$ (is negative in $\alpha_{i_1}...\alpha_{i_n}$, is positive in $\alpha_{i_1}...\alpha_{i_n}$) for some permutation $i_1, ..., i_n$ of $1, ..., n$. Clearly, in the above definitions of legal P, p occurring in P, and p being negative/positive in P, one can change the existential quantifier "for some permutation" into the universal quantifier "for all permutations".

For a string $u = pq$ with $p, q \in \Pi_k$ we say that p is *left in u*, and q is *right in u*. A legal set of strings $P = \{\alpha_1, ..., \alpha_n\}$ is a *disjoint cycle* if there exist pointers $p_1, ..., p_n, p_{n+1}$ such that:
(1) $|\alpha_i| = 2$ for all $1 \leq i \leq n$,
(2) $\alpha_1 = p_1p_2$, and for each $2 \leq i \leq n$, α_i contains one occurrence from $\{p_i, \bar{p}_i\}$ and one occurrence of p_{i+1}; moreover $p_{n+1} \in \{p_1, \bar{p}_1\}$,
(3) for each $2 \leq i \leq n$, if p_i is negative in P, then p_i occurs in both α_{i-1} and α_i; moreover, p_i is left in α_i if and only if p_i is right in α_{i-1},
(4) for each $2 \leq i \leq n$, if p_i is positive in P, then $\bar{p}_i$ occurs in α_i and p_i occurs in α_{i-1}; moreover $\bar{p}_i$ is left in α_i if and only if p_i is left in α_{i-1}.

Let u be a string over Π_k. We say that u *contains* P if each string from P has a cyclic occurrence in u in such a way that these occurrences are pairwise disjoint; otherwise we say that u *avoids* P. In particular, we will be interested in the situations where P is a disjoint cycle, and u is a legal word.

Example 8 *Let* $P = \{53, 34, 45\}$.
Since, e.g., 345345 *is a legal string,* P *is legal. Also* P *is a disjoint cycle, because for* $\alpha_1 = 34, \alpha_2 = 45, \alpha_3 = 53, p_1 = 3, p_2 = 4,$ *and* $p_3 = 5$, *conditions (1) through (4) of the definition of a disjoint cycle are satisfied. Then* $u = 3234\bar{2}4566\bar{6}5$ *contains* P, *because* 34, 45 *and* 53 *are disjoint cyclic substrings of* u *(with* 53 *"split in two" at the end of* u*). On the other hand* $u_1 = \bar{2}3453254$ *avoids* P *(in spite of the fact that each string of* P *is a substring of* u_1*).*

Theorem 7 *A realistic string* u *has a successful reduction in* $DLAD_k$ *if and only if*
(1) u *consists of negative pointers only, and*
(2) u *avoids disjoint cycles.*

Consider the real MDS structure given by the realistic MDS descriptor $\delta = (2,3)(5,e)(3,4)(b,2)(4,5)$, which yields the realistic string $u = 23534245$. Hence, u contains four pointer sets, u consists of negative pointers only, and one can check that u avoids disjoint cycles. Consequently, by Theorem 7, u has a successful reduction in $DLAD_k$ – as a matter of fact, $(u; \mathbf{sdr}_{2,4}, \mathbf{sdr}_{3,5})$ is a successful reduction.
Consider now the real MDS structure given by the realistic MDS descriptor $\delta = (b,2)(4,5)(3,4)(2,3)(5,e)$, which yields the realistic string 24534235. This string u contains a disjoint cycle $P = \{24, 42\}$, and so, by Theorem 7, it does not have a successful reduction in $DLAD_k$.

We will discuss now the combinations of single operations.

Theorem 8 *A realistic string* u *has a successful reduction in* $LD_k \cup HI_k$ *if and only if the following holds for each legal substring* v *of* u*: if* $v = v_1u_1v_2u_2...v_ju_jv_{j+1}$, *where each* u_i *is a legal substring, then* $v_1v_2...v_{j+1}$ *either contains a positive pointer or it is reducible in* LD_k.

Consider the real MDS structure given by the realistic MDS descriptor $\delta = (b,2)(\bar{4},\bar{3})(2,3)(4,5)(5,e)$, which yields the realistic string $u = 2\bar{4}\bar{3}23455$. Since the only legal substrings of u are 55, $2\bar{4}\bar{3}234$, and u itself, it is easily seen that u satisfies the condition from the statement of Theorem 8. Hence, u has a successful reduction in $LD_k \cup HI_k$ – as a matter of fact $(u; \mathbf{spr}_3, \mathbf{spr}_2, \mathbf{spr}_4, \mathbf{snr}_5)$ is a successful reduction.
Consider now the real MDS structure given by the realistic MDS descriptor $\delta = (b,2)(3,4)(\bar{5},\bar{4})(5,6)(2,3)(6,e)$ which yields the realistic string $u = 234\bar{5}\bar{4}56236$. Now, for $v = u$ and $v_1 = 23, u_1 = 4\bar{5}\bar{4}5, v_2 = 6236$, after removing u_1 we get the string $v_1v_2 = 236236$ which does not have a positive pointer,

and it is clearly not reducible in LD_k. Thus, by Theorem 8, u does not have a reduction in $LD_k \cup HI_k$.

Theorem 9 *A realistic string u has a successful reduction in $LD_k \cup DLAD_k$ if and only if all pointers in u are negative.*

Thus, according to Theorem 9, the realistic string $u = 23452345$ (obtained from the realistic MDS descriptor $\delta = (2,3)(4,5)(b,2)(3,4)(5,e)$) has a successful reduction in $LD_k \cup DLAD_k$ – as a matter of fact $(u; \mathbf{sdr}_{3,5}, \mathbf{snr}_4, \mathbf{snr}_2)$ is a successful reduction.
Also, the realistic MDS descriptor δ corresponding to the MDS structure of the α-TP gene in *Oxytricha nova*, see [32], is
$\delta = (b,2)(3,4)(5,6)(7,8)(9,10)(11,12)(2,3)(4,5)(6,7)(8,9)(10,11)(12,13)(13,14)(14,e)$.
The corresponding realistic word is
$u = 2\,3\,4\,5\,6\,7\,8\,9\,10\,11\,12\,2\,3\,4\,5\,6\,7\,8\,9\,10\,11\,12\,13\,13\,14\,14$, and, according to Theorem 9, u has a successful reduction in $LD_k \cup DLAD_k$.

Results characterizing patterns of micronuclear genes for various subsets of $\{ld, hi, dlad\}$ can be used to determine the *operational complexity* and the *operational similarity* of micronuclear genes. Thus, e.g., one can prove that any strategy for the assembly of the actin I gene of *Oxytricha trifallax must* use all three molecular operations, and the same holds for the actin I gene of *Oxytricha nova.* On the other hand the α-TP gene of *Oxytrycha nova* can be successfully assembled using only the *ld* and *dlad* operations (and it cannot be successfully assembled using any other strict subset of our three operations). Finally, the R1 gene of *Oxytrycha nova* can be successfully assembled using only the *ld* operation. We can therefore claim that, from the operational complexity point of view, the R1 gene of *Oxytrycha nova* is the simplest of the four genes listed above, the α-TP gene of *Oxytrycha nova* is more complex, and the actin I genes in *Oxytrycha trafallax* and in *Oxytrycha nova* are the most complex (and similar to each other, because both of them *must* use all three operations in any assembly process).

This section is based on [7].

Discussion

In this tutorial/survey paper we have presented some basic results concerning the computational aspects of gene assembly in ciliates. We believe that the results obtained so far are interesting for both biology and computer science.

From the biological point of view our intramolecular model is interesting because
(i) it provides one *uniform* explanation of *all* known cases of gene assembly in hypotrichous ciliates, and
(ii) it provides a set of *formal tools* to study the process of gene assembly (an example of a formal system for analyzing gene assembly is given in [36]).

From the computer science point of view, the most astounding discovery is that ciliates discovered the *linked list* (one of the fundamental data structures used in computer science see, e.g., [5]) hundreds of millions of years ago (!!!) and implement it through a very elegant pattern matching mechanism. The investigation of computational aspects of gene assembly in ciliates leads to novel models and problems that are interesting for theoretical computer science. In particular it leads to a new paradigm: *computing by folding and recombination*. The investigation of this paradigm will hopefully enrich theoretical computer science, and it may provide a better insight into *natural computing*, i.e., computing that is inspired by (gleaned from) nature.

Because of the limitation on the size of this paper we could not cover all important topics of the current research on gene assembly. Let me mention now some of them.

As discussed in Section 4, the gene assembly may end with a given gene placed in a linear or in a circular molecule. A simple algorithm for deciding which case holds is presented in [12], but, more important, the following is proved there. In general (as discussed already in Section 9), there are many possible strategies for assembling a macronuclear gene from its micronuclear precursor. However, if *any* of these strategies leads to an assembly in a circular molecule, then *all* strategies will lead to an assembly in a circular molecule. In other words: "*Once* circular – *always* circular". This property is thus an *invariant* of gene assembly. Since we do not know (yet) which strategy is used by ciliates in the assembly of a given macronuclear gene, it is very important to study invariants, i.e., the properties that hold independently of the strategy used. The circularity and other invariants are discussed in [12].

The process of gene assembly can be seen as a process of dynamically changing cyclic graph decomposition. The nodes of the underlying graph are either of valency two (as, e.g., markers) and of valency 4 (when two occurrences from a pointer set are "brought together" to form a *pointer node*). Now recombining on a given pointer (e.g., by using one of our three molecular operations) corresponds to a "local rerouting" in the corresponding pointer node. Such a rerouting changes the decomposition of the underlying graph into cycles. This idea is exploited in [9], and the model discussed there is more general than the model of assembly based on $\{ld, hi, dlad\}$. – it also covers the *intermolecular* gene assembly. In fact, the model from [7] can be considered as a general framework for systems based on "pointer-like" recombination.

Some basic properties of the recombination mechanism are considered in [39]. These properties are of a topological nature: e.g., one investigates the connectivity of the result depending on the neighbourhood relationship of the input molecule(s).

Finally, there is a wealth of results in papers dealing with the *intermolecular* model of Landweber and Kari – see [17], [18], [19], [22] and [23]. Obviously, in order to understand the *computing by folding and recombination* paradigm one must necessarily investigate both the intramolecular and the intermolecular models. Therefore, these two lines of research complement each other, and together they form a broad framework for the understanding of this novel and exciting paradigm.

Acknowledgements

I am grateful to Laura Landweber and Lila Kari for bringing to the attention of the DNA computing community the computational beauty of gene assembly in ciliates, and in particular for introducing me to this topic during their visit to Leiden in 1998. I am indebted to Andrzej Ehrenfeucht and David Prescott from Boulder, Colorado, for joining me, from the very beginning, in this exciting "ciliates adventure". Without their brilliant insights into the mathematical and biological aspects of computing by ciliates, the work presented here would never have materialized. Special thanks go to David for his patient and insightful coaching in the biology of ciliates (he really knows "everything" about them). Also, I am very thankful to Tero Harju and Ion Petre from Turku for joining the three of us in forming *our ciliates team.* Their mathematical insights and techniques (and enthusiasm) are very important for our research. Finally, I am grateful to Marloes van der Nat for the expert typing of many iterations of this paper, to Ion Petre for his help in producing Figure 4, and to Hendrik Jan Hoogeboom for comments on the previous version of this paper.

References

[1] Adleman, L.M., Molecular computation of solutions to combinatorial problems, Science, 226, 1021-1024 (1994).

[2] Apostolico, A., String editing and longest common subsequences, in *Handbook of Formal Languages*, Rozenberg, G. and Salomaa, A. (eds.), Springer-Verlag, Berlin, Heidelberg, v. 3, 361-398 (1997).

[3] Berg, P., and Singer, M., *Dealing with Genes: The Language of Heredity*, University Science Books, Mill Valey (1992).

[4] Condon, A., and Rozenberg, G. (eds.), *DNA Computing: Proceedings of the 6th DNA Based Computers meeting, Lecture Notes in Computer Science*, Springer-Verlag, Berlin, Heidelberg, v. 2054 (2001).

[5] Cormen, T.H., Leiserson, C.E., and Rivest, R.L., *Introduction to Algorithms*, MIT Press, Cambridge, London, (1990).

[6] Crochemore, M., and Hancart, C., Automata for matching patterns, in *Handbook of Formal Languages*, Rozenberg, G., and Salomaa, A. (eds.), Springer-Verlag, Berlin, Heidelberg, v. 3, 399-462 (1997).

[7] Ehrenfeucht, A., Harju, T., Petre, I., and Rozenberg, G., Characterizations of patterns of micronuclear genes in ciliates, manuscript (2001).

[8] Ehrenfeucht, A., Harju, T., Petre, I., Prescott, D.M., and Rozenberg, G., Formal systems for gene assembly in ciliates, to appear in Theoretical Computer Science (2001).

[9] Ehrenfeucht, A., Harju, and T., Rozenberg, G., Gene assembly through cyclic graph decomposition, to appear in Theoretical Computer Science (2002).

[10] Ehrenfeucht, A., Petre, I., Prescott, D. M., and Rozenberg, G., Universal and simple operations for gene assembly in ciliates, in *Where Mathematics, Computer Science, Linguistics and Biology Meet*, Martin-Vide, C., and Mitrana, V. (eds.), Kluwer Academic Publishers, Dordrecht/Boston, 329–342 (2001).

[11] Ehrenfeucht, A., Petre, I., Prescott, D.M., and Rozenberg, G., String and graph reduction systems for gene assembly in ciliates, to appear in Mathematical Structures in Computer Science (2001).

[12] Ehrenfeucht, A., Petre, I., Prescott, D.M., and Rozenberg, G., Circularity and other invariants of gene assembly in ciliates, manuscript.

[13] Ehrenfeucht, A., Prescott, D.M., and Rozenberg, G., Computational aspects of gene (un)scrambling in ciliates. In *Evolution as Computation*, L. Landweber, E. Winfree (eds.), 45–86, Springer Verlag, Berlin, Heidelberg (2001).

[14] Engelfriet, J., and Rozenberg, G., Fixed point languages, equality languages, and representations of recursively enumerable languages, Journal of the ACM, 499-518 (1980).

[15] Head, T., Formal language theory and DNA: an analysis of the generative capacity of specific recombinant behaviors, Bulletin of Mathematical Biology, 49, 737-759 (1987).

[16] Head, T., Paun, Gh., and Pixton, D., Language theory and molecular genetics: generative mechanisms suggested by DNA recombination, in *Handbook of Formal Languages*, Rozenberg, G., and Salomaa, A. (eds.), Springer-Verlag, Berlin, Heidelberg, v. 2, 295-360 (1997).

[17] Kari, J., and Kari, L., Context free recombinations, in *Where mathematics, computer science, linguistics and biology meet*, Martin-Vide, C. and Mitrana, V. (eds.), Kluwer Academic Publishers, Dordrecht/Boston, 361-375 (2000).

[18] Kari, L., and Landweber, L.F., Computational power of gene rearrangement, in Winfree, E., and Gifford, D. (eds.), *Proceedings of DNA Based Computers V*, American Mathematical Society, 207–216 (1999).

[19] Kari, L., Kari, J., and Landweber, L.F., Reversible molecular computation in ciliates, in *Jewels are Forever*, Karhumäki, J., Maurer, H., Paun, G., and Rozenberg, G. (eds.), Springer Verlag, Berlin, Heidelberg, 353-363 (1999).

[20] Kleene, S.C., Representation of events in nerve nets and finite automata, in: *Automata Studies*, Shannon, C.E., and McCarthy, J. (eds.), Princeton University Press, 3-41 (1956).

[21] Knigt, T.F., Jr., and Sussman, G.J., Cellular gate technology, in *Unconventional Models of Computations*, Calude, C.S., Casti, J., and Dinneen, M.J. (eds.), Springer Verlag, Singapore (1998).

[22] Landweber, L.F., and Kari, L., The evolution of cellular computing: nature's solution to a computational problem, Biosystems, 52, 3–13 (1999).

[23] Landweber, L.F., and Kari, L., Universal molecular computation in ciliates. In *Evolution as Computation*, Landweber, L., Winfree, E. (eds.), Springer Verlag, Berlin, Heidelberg, to appear (2001).

[24] Lindenmayer, A., Mathematical models for cellular interaction in development, I and II, Journal of Theoretical Biology, 18, 280-315 (1968).

[25] McCulloch, W.S., and Pitts, W., A logical calculus of the ideas imminent in nervous activity, Bulletin of the Mathematical Biophysics, 5, 115-133 (1943).

[26] Myers, G., Whole-genome DNA sequencing, *Computing in Science and Engineering*, 1, 33-43 (1999).

[27] von Neumann, J., *Theory of Selfreproducing Automata* (ed. by Burks, A.W.), University of Illinois, Urbana (1966).

[28] Paun, Gh., Computing with membranes, Journal of Computer and System Sciences, 61, 108-143 (2000).

[29] Paun, Gh., and Rozenberg, G., A guide to membrane computing, Theoretical Computer Science, to appear.

[30] Paun, Gh., Rozenberg, G., and Salomaa, A., *DNA Computing*, Springer-Verlag, Berlin, Heidelberg (1998).

[31] Pevzner, P.A., *Computational Molecular Biology: An Algorithmic Approach*, MIT Press, Cambridge, Massachusetts (2000).

[32] Prescott, D.M., Cutting, splicing, reordering, and elimination of DNA sequences in Hypotrichous ciliates. BioEssays, 14 (5): 317–324 (1992).

[33] Prescott, D.M., The unusual organization and processing of genomic DNA in Hypotrichous ciliates. Trends in Genetics 8, 439–445 (1992).

[34] Prescott, D.M., Genome gymnastics: unique modes of DNA evolution and processing in ciliates, Nature Reviews in Genetics, 1, 191-198 (2000).

[35] Prescott, D.M., and DuBois, M., Internal eliminated segments (IESs) of Oxytrichidae, Journal of Eukariotic Microbiology 43, 432–441 (1996).

[36] Prescott, D.M., Ehrenfeucht, A., and Rozenberg, G., Molecular operations for DNA processing in Hypotrichous ciliates, to appear in European Journal of Protistology (2001).

[37] Prusinkiewicz, P., and Lindenmayer, A., *The Algorithmic Beauty of Plants*, Springer-Verlag, New York (1990).

[38] Rozenberg, G., and Salomaa, A., *The Mathematical Theory of L Systems*, Academic Press, New York (1980).

[39] Rozenberg, G., and Sant, P., Safe recombination, manuscript (2001).

[40] Salomaa, A., *Jewels of Formal Language Theory*, Computer Science Press, Rockville, Md. (1981).

[41] Smith, W.D., DNA computers *in vitro* and *in vivo*, in *DNA Based Computers: Proceedings of a DIMACS Workshop*, Lipton, R.J., and Baum, E.B. (eds.), American Mathematical Society (1996).

[42] Weiss, R., and Knight, T.F., Jr., Towards *in vivo* digital circuits, in *Proceedings of the DIMACS Workshop on Evolution as Computation*, Princeton, 1–18, 1999.

COMPOSITIONS OVER A FINITE DOMAIN: FROM COMPLETENESS TO SYNCHRONIZABLE AUTOMATA

ARTO SALOMAA
Turku Centre for Computer Science
Lemminkäisenkatu 14A
20520 Turku, Finland
asalomaa@utu.fi

We will consider functions whose domain is a fixed finite set N with n elements, $n \geq 2$, and whose range is included in N. Such a setup occurs in many and very diverse situations. Depending on the interpretation, different questions will be asked. The two interpretations we have had mostly in mind are *many-valued logic* and *finite deterministic automata.* In the former, the set N consists of n *truth values*, and the functions are truth functions. In the latter, the set N consists of the *states* of an automaton, whereas each letter of the input alphabet induces a specific function: the next state when reading that letter. We will consider two specific issues concerning functions of the kind mentioned: *completeness* and *complexity of compositions.* While the former is fairly well understood, very little is known about the latter. Our starting point is an old conjecture, falling within the framework of the complexity of computations, concerning finite deterministic automata. Variants and generalizations of this conjecture are presented. It is also shown that the conjecture does not hold for functions of several variables.

1 Introduction

The classical paper by E.F. Moore, [5], about *Gedanken experiments* on finite automata, had the general idea to view a finite automaton as a black box and to try to find out some specific facts about it by observing what kind of outputs certain inputs produced. Of course, for each experiment, the overall setup has to be defined explicitly. Suppose you know the structure (graph, transition function) of a given finite deterministic automaton A, but do not know the state A is in. How can you get the situation under control? For some automata, not always, there are words, referred to as *synchronizing*, bringing the automaton always to the same state q, no matter from which state you started from. Thus, you first have to feed A a synchronizing word, after which you have the situation completely under control. You can also view the graph of an automaton as a labyrinth, where you are lost. If you then follow the letters of a synchronizing word (and have the global knowledge of the graph of the automaton), you have found your way. This shows the connection with the well-known *road coloring* problem.

Consider now functions as described in the Abstract. The notations N

and n remain fixed throughout the paper but later on we consider also functions of *several* variables ranging over N. A synchronizing word can be viewed as a *composition sequence for a constant.* Specifically, this is done as follows. Consider a finite deterministic automaton, without initial and final states, as a pair $(N, \mathbf{F})$, where N is the state set of cardinality n and $\mathbf{F}$ is a set of functions mapping N into N. The set $\mathbf{F}$ determines both the input alphabet and the transition function in the natural way, and input words correspond to compositions of functions. We can read compositions from left to right to be in accordance with the customary way of reading input words from left to right. In the sequel it will always be clear from the context in which direction we are reading compositions: $g(f(x))$ or $((x)g)f$.

An automaton is *synchronizable* if and only if it possesses a synchronizing word. This happens exactly in case a constant function is in $\mathbf{G(F)}$, the set of all functions generated by $\mathbf{F}$ by compositions. The *Černý Conjecture* says that every synchronizable automaton possesses a synchronizing word of length $\leq (n-1)^2$. The conjecture was originally presented in [2] and discussed further, for instance, in [3]. It became well known through the work J.-E. Pin, [7]. Numerous papers have appeared about this topic since 1995, see [1] for a survey. Overall discussions about the problem, its variants and generalizations are contained also in [4,10,11]. The motivation can be summarized as follows:

- It is an old unsolved problem about the very basics of finite automata.
- The problem fits in the framework of compositions of functions over a finite domain.
- This is a very basic class of functions, coming up in automata, many-valued logic, combinatorics, ...
- Some parts of the theory of such functions are well understood, for instance, *completeness*, some others not *(minimal length of composition sequences)*.
- The problem is closely related to various other problems (*road coloring, experiments with automata, structural theory of many-valued truth functions*).

The next section presents a more general framework (*depth, complete depth*) for the study of problems concerning lengths of composition sequences. Whereas *completeness* is a widely studied and quite well understood notion in many-valued logic, practically nothing is known about the corresponding notion of *complete depth.* An analogous phenomenon can be observed in group theory. While it is pretty well known when two permutations form a *basis* of the symmetric group (see [6]), very little can be said about how long the shortest composition sequence for a specific permutation will be in terms of a given basis. Finding the shortest solution for Rubik's Cube is also a problem about minimizing composition sequences!

Section 3 deals with the depth of constant functions. Sections 4 and 5 extend the theory to concern functions of several variables. It turns out in Section 5 that, for this extension, Černý Conjecture is no more valid.

2 Depth and complete depth

Consider functions $f(x_1, \ldots, x_k)$ as already described in the Abstract. Their variables, finite in number, range over a fixed finite set N with n elements, and their function values are also in N. The notation N, n is fixed throughout the paper, and we also assume $n \geq 3$ to exclude special cases. When we speak of "functions", we always mean functions of this kind. To each function is associated a *positive* integer k, referred to as its *arity.* This section deals with functions of arity 1 (also referred to as *one-place or unary functions*). However, the basic definitions about composition sequences and their length are formulated having in mind also the following sections.

When defining composition sequences, we start with a given finite set **F** of functions. Intuitively, a *composition sequence* is a well-formed sequence of function symbols (letters) from **F** and variables. Its *length* is the maximal amount of nested function symbols in the sequence. The *variables* come from a fixed denumerably infinite collection.

Thus, let **F** be a given finite (nonempty) set of functions. A *composition sequence* (over **F**) and its *length* are defined inductively as follows. A variable alone constitutes a composition sequence of length 0. Assume that f of arity k (≥ 1) is in **F**, and $f_1, \ldots, f_k$ are composition sequences of lengths r_i, $1 \leq i \leq k$. Then $f(f_1, \ldots, f_k)$ is a composition sequence of length $\max\{r_i | 1 \leq i \leq k\} + 1$. Nothing is a composition sequence, unless its being so follows by the above rules.

For instance, if **F** consists of two functions $f(x, y)$ and $g(x)$, then

$$g(f(f(x, g(x)), f(f(x, y), z)))$$

is a composition sequence of length 4. Thus, using a binary function, we obtain composition sequences with arbitrarily many variables. However, in this section we discuss only unary functions. Then the length of a composition sequence equals the length of the word over the "function alphabet" defining the sequence, and only unary functions will result from composition sequences.

Clearly, every composition sequence defines a function, the formal definition of the semantics being obvious. We denote by $\mathbf{G(F)}$ the set of all functions defined by composition sequences over $\mathbf{F}$ with length at least 1. (Thus, the identity function is not automatically in $\mathbf{G(F)}$; it is included only in case it results as a composition.) If we are dealing with unary functions only, then $\mathbf{G(F)}$ can be viewed as the free semigroup generated by $\mathbf{F}$.

For each $k \geq 1$, there are altogether n^{n^k} functions with k variables. A set $\mathbf{F}$ is termed *complete* if all functions, independently of the arity, are in $\mathbf{G(F)}$. It was shown by Post already in the 20's, using a simple minimax argument, that $\mathbf{F}$ is complete if all binary functions are in $\mathbf{G(F)}$. Since unary functions generate (via composition sequences) only unary functions, a set $\mathbf{F}$ of unary functions is termed *complete* if all of the n^n unary functions are in $\mathbf{G(F)}$. For the proof of the following criterion, the reader is referred to [10].

Theorem 1 *Assume that $n \geq 3$. Then three functions generate all functions if and only if two of them generate the symmetric group S_n and the third assumes exactly $n-1$ values. No less than three functions generate all functions.*

We are now ready for the fundamental definitions. Although we consider first only unary functions, the definitions are formulated also with the later sections in mind. The notions were first defined in [4] (where "depth" was referred to as "complexity") and [10], the extension to functions of several variables being due to [11].

Consider a set $\mathbf{F}$ and a function f. The *depth* of f with respect to $\mathbf{F}$, in symbols $D(\mathbf{F}, f)$, is defined as follows. If f is in $\mathbf{G(F)}$, then $D(\mathbf{F}, f)$ equals the length of the *shortest* composition sequence defining f. Otherwise, $D(\mathbf{F}, f) = \infty$.

The defined notion of depth is *relative* to a given set $\mathbf{F}$. We now come to the *absolute* notion of depth. The emphasis will be on unary functions $f(x)$ (because it seems all interesting cases are already thereby exhausted). However, in the first part of the definition, the function f is quite general.

The *depth* $D(f)$ of a function f is defined by the equation

$$D(f) = \max(D(\mathbf{F}, f)), \tag{1}$$

where $\mathbf{F}$ ranges over all sets with the property that f is in $\mathbf{G(F)}$. The *unary depth* $D'(f)$ of a unary function f is defined by the same equation, where now $\mathbf{F}$ satisfies the additional requirement of consisting of only unary functions. Clearly, $D'(f) \leq D(f)$ holds for all unary functions f. Because, for any f, there are sets $\mathbf{F}$ with the specified property, we conclude that the depth of a function is always a positive integer.

The *complete depth* $D_C(f)$ of a function f is defined also as in (1):

$$D_C(f) = \max(D(\mathbf{F}, f)) \tag{2}$$

but now $\mathbf{F}$ ranges over *complete* sets. The further extension to *complete unary depth* is obvious: for a unary function f, the complete unary depth $D'_C(f)$ is defined by the right side of (2), where now $\mathbf{F}$ ranges over complete sets of unary functions.

The following relations are immediate by the definitions.

Lemma 1 *The relation $D_C(f) \leq D(f)$ holds for all functions, and the relations $D'_C(f) \leq D_C(f)$ and $D'_C(f) \leq D'(f) \leq D(f)$ hold for all unary functions.*

Very little is known about the cases where the inequalities are strict. Also the interrelation between $D'(f)$ and $D_C(f)$ is unknown for unary functions f. As was already mentioned, completeness is fairly well understood (see, for instance, [8,9]), whereas very little is known about the depth and its relation to complete depth.

Lemma 2 *For unary functions f, we have $D'(f) \leq n^n$.*

The lemma follows immediately because, in a composition sequence, we can remove any part where the function is not changed. For a proof of the following result, see [10]. Many variants of the result appear in the literature, it can also be stated for some specific functions f.

Theorem 2 *There is no polynomial $P(n)$ such that $D'(f) \leq P(n)$ holds for all unary functions f.*

3 Depth of constants

We assume that our basic set N consists of the first n integers:

$$N = \{1, 2, \ldots, n\}.$$

Among the unary functions we are especially interested in *constants*

$$c_i(x) = i, \text{ for all } x \text{ and } i = 1, 2, \ldots, n.$$

We use the notation SYNCHRO for the class of all sets $\mathbf{F}$ such that at least one of the constants c_i is in $\mathbf{G(F)}$. (In defining SYNCHRO we have some fixed n in mind. Thus, SYNCHRO actually depends on n.) Analogously to the definitions above, we now define the depths

$$D(\text{const}),\ D'(\text{const}),\ D_C(\text{const}),\ D'_C(\text{const}).$$

By definition,

$$D(\text{const}) = \max_{\mathbf{F}} \min\{D(\mathbf{F}, c_i) | 1 \le i \le n\},$$

where **F** ranges over SYNCHRO. The other three depths are defined in exactly the same way, except that **F** ranges over sets of unary functions in SYNCHRO, over complete sets, and over complete sets of unary functions, respectively. Thus, the notions introduced do not deal with the depth of an individual function but rather give the smallest depth of a constant, among the constants generated. Similarly as SYNCHRO, the notions depend on n.

The Černý Conjecture can now be expressed in the following form.

Conjecture 1 (Černý) $D'(\text{const}) = (n-1)^2$.

The finite deterministic automaton defined by the following table has the state set N, and two input letters a (affects a circular permutation) and b (identity except sends n to 1).

δ	1	2	...	$n-1$	n
a	2	3	...	n	1
b	1	2	...	$n-1$	1

The word $(ba^{n-1})^{n-2}b$ is synchronizing and, moreover, there are no shorter synchronizing words and this word is the only synchronizing word of length $(n-1)^2$. We still depict the automaton for $n = 4$.

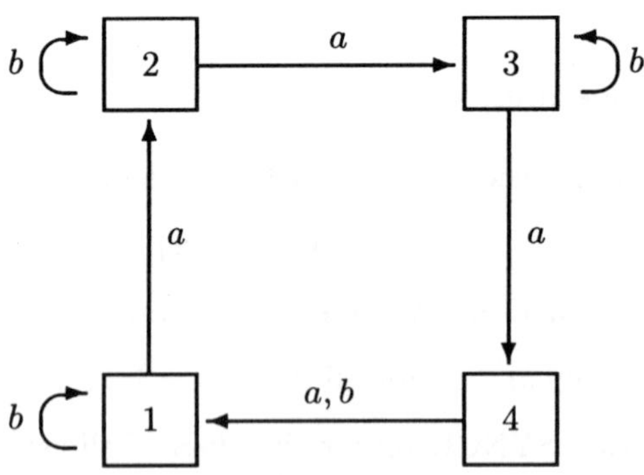

Let **F** consist of the functions a and b defining this particular automaton, and consider the constant functions c_i, $1 \le i \le n$.

Lemma 3 $D(\mathbf{F}, c_i) = (n-1)^2 + i - 1$, *for* $i = 1, 2, \ldots, n$.

See [10] for a proof of the lemma. The lemma yields the lower bound in the following theorem. The reference [10] contains also a proof of the upper bound; various cubic upper bounds are known in the literature for different setups.

Theorem 3 *For a constant* c_i, *we have* $n(n-1) \leq D'(c_i) < n^3/2$. *Moreover,* $D'(\text{const}) \geq (n-1)^2$.

Thus, Conjecture 1 is actually open only as regards the upper bound. On the other hand, the automaton defined above is the only automaton (when isomorphic variants are disregarded) known to us which is defined for an arbitrary n and which reaches the lower bound. Thus, $(n-1)^2$ seems to be a rare exception for the length of the minimal synchronizing word for a finite deterministic automaton, the length being "normally" smaller. This supports Conjecture 1. On the other hand, we will see that if D' is replaced by D, then the statement is not valid.

The variant of Conjecture 1, dealing with the depths of individual functions, can be expressed as follows.

Conjecture 2 *If* c_i *is a constant, then* $D'(c_i) = n(n-1)$.

It is a consequence of Theorem 3 that the depth $n(n-1)$ cannot be reduced. If Conjecure 1 holds, so does Conjecture 2. On the other hand, it is conceivable (although very unlikely) that Conjecture 2 holds but Conjecture 1 fails.

It is not known whether the upper bound ($(n-1)^2$ or $n(n-1)$) is reached for the complete depth. Thus, it is possible that

$$D'_C(\text{const}) < (n-1)^2,$$

and $D'_C(c_i) < n(n-1)$ holds for the constant c_i.

4 Functions of several variables

Binary functions $f(x, y)$ are conveniently defined by tables, where the values of x are read from the rows and those of y from the columns. For instance, consider the *Lukasiewicz implication* $f(x, y)$ (or xyf) defined by $f(x, y) = \max(1, 1 - x + y)$. For $n = 6$, its table is

$$\begin{array}{|cccccc}\hline 1&2&3&4&5&6\\ 1&1&2&3&4&5\\ 1&1&1&2&3&4\\ 1&1&1&1&2&3\\ 1&1&1&1&1&2\\ 1&1&1&1&1&1\end{array}$$

As regards completeness, there is a remarkable difference with respect to Theorem 1: single functions may constitute complete sets. Such functions are referred to as *Sheffer functions*, analogously to the *Sheffer stroke* in propositional logic. The following functions are examples of Sheffer functions in cases $n = 3, 4, 5, 6$.

$$\begin{array}{|ccc}\hline 2&2&*\\ *&3&1\\ 3&*&1\end{array} \qquad \begin{array}{|cccc}\hline 2&2&*&*\\ *&3&1&*\\ *&*&4&3\\ 4&*&*&1\end{array} \qquad \begin{array}{|ccccc}\hline 2&2&*&*&*\\ *&3&1&*&*\\ *&*&4&3&*\\ *&*&*&5&4\\ 5&*&*&*&1\end{array}$$

$$\begin{array}{|cccccc}\hline 2&2&*&*&*&*\\ *&3&1&*&*&*\\ *&*&4&3&*&*\\ *&*&*&5&4&*\\ *&*&*&*&6&5\\ 6&*&*&*&*&1\end{array}$$

The positions marked by the star * can be filled arbitrarily. That the functions obtained are Sheffer functions follows by the criteria in [8,9].

We now consider an example which shows that, for $n = 3$, we have $D(\text{const}) > (n-1)^2$ and even $D_C(\text{const}) > (n-1)^2$. Results for a general n will be obtained in the next section. We consider now the depths $D(\mathbf{F}, g)$, where $\mathbf{F}$ consists of one function only, namely, the Sheffer function $f(x, y)$ defined by the table

$$\begin{array}{|ccc}\hline 2&2&1\\ 1&3&3\\ 1&2&1\end{array}$$

(We have chosen a Sheffer function different from the ones presented above.) Since $\mathbf{F}$ is complete we conclude that, for any function g,

$$D_C(g) \geq D(\mathbf{F}, g).$$

We now determine the depths of all 27 functions of one variable. Each function is represented by its value sequence. Thus, 231 is the circular permutation (123) which is obtained as $f(x,x)$. Since

$$f(f(x,x), f(x,x)) = 312, \;\; f(x, f(x,x)) = 231, \;\; f(f(x,x), x) = 121,$$

we know all functions with depth 1 or 2. The following table, based on an exhaustive search, gives all functions with depth at most 3. The sources are ordered pairs: the target results by applying f to the elements of the pairs.

target	source	depth
231	123, 123	1
312	231, 231	2
121	231, 123	2
123	312, 312	3
232	121, 121	3
112	123, 312	3
122	231, 121	3
211	312, 231	3

We see, for instance, that the function with the value sequence 211 is obtained as $f(f(f(x,x), f(x,x)), f(x,x))$.

It can be shown already now that each of the constants 111, 222, and 333 must be of depth at least 5. Since the 3's in the definition of f are in the second row, the constant 222 is the necessary first component in the source of 333. The above table shows that all functions of depth at most 3 have the number 2 in their value sequence. On the other hand, since all 1's (resp. 2's) in the definition of f are in the first and third columns (resp. rows), the second (resp. first) component in the source of 111 (resp. 222) cannot have the number 2 in its value sequence and, hence, must be of depth at least 4, which shows that all constants are of depth at least 5.

Continuing the table, we obtain the functions of depths 4 and 5:

target	source	depth		target	source	depth
313	232, 232	4	•	331	223, 223	5
223	112, 112	4	•	133	322, 322	5
233	122, 122	4	•	332	221, 221	5
322	211, 211	4	•	323	212, 212	5
311	232, 231	4	•	321	231, 223	5
213	312, 232	4	•	111	231, 113	5
113	122, 312	4	•	131	223, 121	5
221	312, 211	4	•	132	121, 322	5
212	121, 211	4	•	222	113, 212	5

So far we have 26 functions and are still missing the constant 333. We already noticed that since the 3's in the definition of f are in the second row, the constant 222 is the necessary first component in the source of 333. (The second component must avoid the value 1; 232 is the function with the smallest depth having this property.) Consequently, 333 has the source (222,232) and is of depth 6. Thus, all constants 111, 222, and 3333 are of depth 5 or 6.

5 Forcing long composition sequences

We denote by $\mathbf{F}'$ the set consisting of two functions $g(x)$ and $f(x,y)$, where $g(x)$ is the circular permutation $(1, 2, \ldots, n)$ and $f(x, y)$ is defined by

$$f(x, y) = x \text{ except } f(n, 1) = 1.$$

Thus, for $n = 4$, the value sequence of g is 2341, and $f(x, y)$ is defined by the table

1	1	1	1
2	2	2	2
3	3	3	3
1	4	4	4

Lemma 4 *All constants c_i satisfy $D(\mathbf{F}', c_i) > (n-1)^2$. The constant c_n satisfies $D(\mathbf{F}', c_n) > n(n-1)$.*

Proof. We use the term *(S)-function* for functions assuming only values in the set S. Thus, a (1,2,4)-function assumes only values 1,2 and 4 (perhaps not all of them). When we speak about composition sequences, we mean composition sequences over $\mathbf{F}'$. Let now α be the *shortest* composition sequence for a constant. (If there are several shortest sequences, we fix one of them.) We cannot have $\alpha = g(X)$, for some composition sequence X, because then also X

defines a constant and is shorter. Consequently, we must have $\alpha = f(f_1, f_2)$, for some f_1 and f_2. We analyze f_1, ignoring f_2. Recall that our aim is to establish the lower bound $(n-1)^2+1$ for the length of α. Since f_1 cannot be a constant (we would again have a shorter composition sequence), f_1 must be a (1,n)-function. This follows by the definition of f. We also infer that c_1 is the constant defined by α.

Let f_1' be the shortest composition sequence for a (1,n)-function. Clearly,

$$|\alpha| \geq |f(f_1', f_2)|.$$

We continue by analyzing f_1'. Because f_1' is a (1,n)-function, we cannot have $f_1' = f(f_3, f_4)$, since then f_3 would be either a (1,n)-function or the constant c_n. Both alternatives are impossible by our length assumptions. Hence, $f_1' = g(g_1)$, which implies that g_1 is an ($n-1$,n)-function. We now analyze g_1, first replacing it by an ($n-1$,n)-function with the shortest composition sequence. We infer that the composition sequence of g_1 must begin with g. Continuing in the same way, we conclude that $f_1' = g^{n-1}g_{n-1}$, where g_{n-1} is a (1,2)-function. Another application of g would lead to another (1,n)-function, which would contradict again our length assumptions. Consequently, we must have $g_{n-1} = f(f_1'', f_2'')$. By the definition of f and our length assumptions, we conclude that f_1'' is a (1,2,n)-function.

Continuing in this way, we reach the composition sequence $(fg^{n-1})^{n-2}h$, where h is a $(1, 2, \ldots, n-1)$-function. (We have ignored parentheses in the composition sequence.) Obviously, $f(x, g(x))$ is the shortest composition sequence for such an h. Thus, we conclude that

$$|\alpha| \geq n(n-2) + 2 > (n-1)^2.$$

Since no other constants are generated before c_1 and it takes $n-1$ steps to get c_n from c_1, we get the lower bound for the depth of c_n as claimed in the lemma.

To complete the proof, we still have to show that $\mathbf{F}'$ is in SYNCHRO. (The lower bounds do not exclude the case that the depths we are interested in are infinite!) But this can be easily accomplished inductively. We give the explicit construction for $n = 4$. Denote

$$r(x) = gggf(x, g(x)) \text{ and } s(x) = f(r(x), f(x, g(x))).$$

Then the constant c_1 is defined by $f(gggs(x), s(x))$. •

Our final result is an immediate consequence of the lemma. The result should be contrasted to Conjectures 1 and 2.

Theorem 4 *For a constant* c_i *we have* $D(c_i) > n(n-1)$. *Moreover,* $D(\text{const}) > (n-1)^2$.

Many-valued functional constructions offer rich possibilities for "forcing" long composition sequences, as was done in the proof above. We mention, finally, an example of a different kind. Assume that $n = 3$ and let $\mathbf{F}$ consist of three functions: transpositions (12) and (23), and the binary function $f(x, y)$ defined by

$$f(x, y) = x \text{ except } f(1, 3) = 2.$$

One can then show by a direct forcing argument that, although $\mathbf{F}$ is complete, no constant has a composition sequence of length smaller than 5.

6 Conclusion

The notion of depth constitutes a very natural aspect of the complexity of a given function with respect to a specific set of functions. We have introduced several variants of this notion, linking it also together with an old unsolved problem about synchronization in finite deterministic automata. While very little is known about these notions of depth, also synchronizability in general remains to be characterized: no conditions characterizing sets in SYNCHRO are known. Using properties of *self-conjugate* functions (see [10,4]), good examples of sets not in SYNCHRO can be constructed.

References

1. S. Bogdanović, B. Imreh, M. Ćirić and T. Petković, Directable automata and their generalizations: a survey. Novi Sad J. Math. 29 (1999).
2. J. Černý, Poznámka k homogénnym experimentom s konečnými automatmi. Mat.fyz.čas SAV 14 (1964) 208-215.
3. J. Černý, A. Pirická and B. Rosenauerová, On directable automata. Kybernetika 7 (1971) 289-298.
4. A. Mateescu and A. Salomaa, Many-valued truth functions, Černý's conjecture and road coloring. EATCS Bulletin 68 (1999) 134-150.
5. E.F. Moore, Gedanken experiments on sequential machines. In C.E. Shannon and J. McCarthy (ed.) Automata Studies, Princeton University Press (1956) 129-153.
6. S. Piccard, Sur les bases du groupe symétrique et les couples de substitutions qui engendrent un goupe régulier. Librairie Vuibert, Paris (1946).
7. J.-E. Pin, Le problème de la synchronisation, Contribution a l'étude de la conjecture de Černý. Thèse de 3e cycle a l'Université Pierre et Marie Curie (Paris 6) (1978).

8. A. Salomaa, A theorem concerning the composition of functions of several variables ranging over a finite set. J. Symb. Logic 25 (1960) 203-208.
9. A. Salomaa, On basic groups for the set of functions over a finite domain. Ann. Acad. Scient. Fennicae, Ser. AI, 338 (1963).
10. A. Salomaa, Composition sequences for functions over a finite domain. Turku Centre for Computer Science Technical Report 332 (2000).
11. A. Salomaa, Depth of functional compositions. EATCS Bulletin 71 (2000) 143-150.

Appendix

Brief Biographies of the Authors

JANUSZ (JOHN) A. BRZOZOWSKI

Janusz (Ya-noosh), or John, Brzozowski was born in Warsaw, Poland and received his BASc and MASc (1957, 1959) from the University of Toronto and an MA and PhD from Princeton University (1962). He spent many years as Professor in the Department of Computer Science at the University of Waterloo, including two periods as the Chair of the Department. He is currently a Distinguished Professor Emeritus in the Computer Science Department of the University of Waterloo and has had visiting appointments at various Universities throughout the world, including University of California, Berkeley and the University of Paris.

Dr. Brzozowski has authoured or co-authoured over 50 papers published in a variety of journals, including Theoretical Computer Science and Discrete Applied Mathematics as well as over 50 conference papers. He has also contributed to 7 books and written two, including 'Asynchronous Circuits' (1995).

He has a hobby in natural languages. When visiting another country X, he learns to say "I speak X-ish very well, but I don't understand it at all" and sound like a native X-ish dweller. This results in a flood of X-ish in his general direction, since no one believes it.

It also turns out, according to his research in the 1960's, that all the misspellings of his last name is not regular - it is not even context-free.

His current research includes Asynchronous Circuits, Testing, Models and Algorithms related to VSLI and Automata and Formal Languages.

JURIS HARTMANIS

Born in Riga, Latvia, on July 5, 1928, Hartmanis received his undergraduate degree in physics from the University of Marburg, a master's degree in mathematics from the University of Kansas City, and a Ph. D. in mathematics from the California Institute of Technology in 1955.

After graduating from Caltech, Hartmanis served on the mathematics faculties at Cornell and Ohio State Universities and then as a research mathematician at the General Electric Research Labs. In 1965 he joined (as chair) the newly-formed computer science department at Cornell University, and became the Walter R. Read Professor of Engineering in 1980. From 1996 to 1998 Hartmanis held the position of Assistant Director of the National Science Foundation for the Directorate of Computer and Information Science and Engineering and returned to Cornell in 1999.

Most of his research has been in automata theory and computational complexity. In 1993, Hartmanis and R.E. Stearns were awarded the ACM Turing Award "in recognition of their seminal paper which established the foundations for the field of computational complexity theory". He has received several other awards, including the Bolzano Gold Medal of the Academy of Sciences (Czech Republic) in 1995, Honorary Doctorate Degrees from University of Dortmund in 1995 and from the University of Missouri in 1999, the Humboldt Foundation Senior US Scientist Award and very recently the Computing Research Association's 2000 Distinguished Service Award. He is a member of the National Academy of Engineering, American Academy of Arts and Sciences, Latvian Academy of Sciences and New York State Academy of Sciences.

Hartmanis has also served on the Turing Award Committee, the Gödel Prize Committee and the Waterman Award Committee; has edited for Springer-Verlag Lecture Notes in Computer Science, Journal of Computer and Systems Sciences, Fundamenta Informaticae, and SIAM Journal on Computing. He serves on the advisory board for EATCS Monographs in Theoretical Computer Science (Springer-Verlag) and the editorial boards for Chicago Journal of Theoretical Computer Science, Electronic Journal for the Foundation of Computer Science, and Journal of Universal Computer Science.

JOHN E. HOPCROFT

After receiving his Ph.D. in electrical engineering from Stanford University in 1964, Professor John E. Hopcroft spent three years on the faculty of Princeton University. He joined the Cornell University faculty in 1967, where he served as chairman of the Department of Computer Science from 1987 to 1992. In 1990, he received an honorary Doctor of Humanities Degree from the University of Seattle. From 1992-1993 he served as Associate Dean for College Affairs in the College of Engineering at Cornell, and was appointed the Joseph Silbert Dean of the College of Engineering in 1994 and continues his work there today. He currently oversees ten academic departments as well as various research units. He has not abandoned the Computer Science department, however, but continues his involvement through his research on robust geometric algorithms, modeling and simuation, and information capture and access. He is also working closely with the Design Research Institute in developing technologies to facilitate information capture and access within an engineering design environment.

Through his pioneering work in the theoretical aspects of computing, namely algorithm analysis, formal languages, automata theory, and graph theory, Professor Hopcroft has earned the recognition which he has received and which he so deserves. His books, co-authored with Professor J.D. Ullman and Dr. A.V. Aho, are classics in the field. His great contributions in theoretical computer science have awarded him the 1986 Turing Award–for fundamental achievements in the design and analysis of algorithms and data structures–along with fellowhips in the American Academy of Arts and Sciences, the Insitute of Electrical and Electronics Engineers, and the National Academy of Engineering. He was even appointed by President Bush to the National Science Foundation's National Science Board in 1992, where he served until 1998.

He is currently a member of the Association for Computing Machinery, the Society for Industrial and Applied Mathematics, the American Association for the Advancement of Science, the New York Academy of Science, and the Institute of Electrical and Electronics Engineers. He is a Fellow of the American Academy of Arts and Sciences, the American Association for the Advancement of Science, the Institute of Electrical and Electronics Engineers, and the Association for Computing Machinery. He is chairman of the SIAM Board of Trustees, member of the Scientific Advisory Committee for the David and Lucile Packard Foundation, and of the Sloan Research Fellowship Committee. Also, he is on the advisory board for the Supercomputing Research

Center, Institute for Defense Analysis. He is editor of the Oxford University Press International Series on Computer Science, Algorithmica, Discrete and Computational Geometry, and associate editor of the International Journal of Computational Geometry and Applications, the Journal of Computer and Systems Sciences, and the Journal of Information Sciences.

WERNER KUICH

Born in 1941, Werner Kuich studied in Wien. In his youth he worked for some years at IBM although most of his career he has worked as a professor at the Technical University of Wien. Dr. Kuich is an expert on formal power series and semirings, and in addition to publishing many papers in various well-known Journals, he has written a chapter entitled “Semirings and Formal Power Series” in the “Handbook of Formal Languages”, and a book entitled “Semirings, Automata, Languages” with Dr. Arto Salomaa. All of his work is characterized by accuracy and rigor.

Dr. Kuich enjoys travelling, and has spent a time in various European countries and a year in North America. He was a well-known bachelor in the theoretical computer science community until he got married in his late thirties. He has a beautiful house in Brunn near Wien, where he lives with his wife Beate and his cat. Werner is one of the founding fathers of treating automata and languages in terms of power series.

In addition to the amazing work accomplished in theoretical Computer Science Dr. Kuich has also written some beautiful sauna poetry which can be read in the book “TCS, People and Ideas”.

ROBERT MCNAUGHTON

Dr. McNaughton is currently Emeritus Professor of Computer Science at Rensselaer Polytech Institute. He was educated at Columbia University where he received his BA in 1948, completing his studies at Harvard University in 1951 when he received his PhD. Awards received include the Levy Medal in 1956.

Robert McNaughton entered Computer Science in the 1950s after teaching philosophy for six years, a career switch which was due to the lean job market more than anything else. Today, however, his training in philosophy holds him in good stead.

McNaughton, who is author of the textbook "Elementary Computability, Formal Languages and Automata" published by Prentice-Hall, is now looking at problems in the combinatorics of words, a branch of formal languages. Currently Emeritus Professor of Mathematics and computer science at Rensselaer Polytechnic Institute, Dr. McNaughton's research is being coordinated with computer scientists both in the Computer Science Department at Rensselaer, and at the University of Albany. Their research is concerned with looking at formal linguistic systems for the sake of carrying through proofs on a machine. For example, they have looked at ways to improve the efficiency of Thue systems, a linguistic method developed by Norwegian logician Axel Thue in 1914. Thue systems are useful for computation because they replace strings (connected characters) with other strings, carrying through a rather basic kind of computer operation.

GRZEGORZ ROZENBERG

In the words of his son at age 14, Grzegorz Rozenberg "is a university professor", to which he would add after a short pause, "but he is not stupid, he is a very good magician."

Professor Rozenberg received his Master and Engineering degree in computer Science in 1965 from the Technical University of Warsaw, Poland. In 1968 he obtained his Ph.D. in mathematics at the Polish Academy of Sciences, Warsaw. Since then he has held full time positions at the Polish Academy of Sciences, Warsaw, Poland, Utrecht University, The Netherlands, State University of New York at Buffalo, U.S.A., and University of Antwerp, Belgium. Since 1979 he has been a professor at the Department of Computer Science of Leiden University and an adjoint professor at the Department of Computer Science of University of Colorado at Boulder, U.S.A. He is the head of the Theoretical Computer Science group at Leiden Institute of Advanced Computer Science, and the scientific director of Leiden Center for Natural Computing. Professor Rozenberg has published about 350 papers, 5 books, and is an editor/co-editor of about 50 books. He is a Foreign Member of the Finnish Academy of Sciences and Letters, a member of Academia Europaea, and a doctor honoris causa of the University of Turku, Finland.

From 1985 to 1994, Professor Rozenberg was the President of the European Association for Theoretical Computer Science, and was the chairman of the Award Committee for the Gdel Prize in 1997. His current functions include:

- Chairman of the Steering Committee for International Conferences on Theory and Applications of Petri Nets,
- Chairman of the Steering Committee for International Conferences on Developments in Language Theory,
- Chairman of the Steering Committee for DNA Based Computers Conferences,
- Vice Chairman of the Steering Committee for the International Conferences on Graph Transformation,
- Chairman of the Steering Committee for European Educational Forum,
- Director of the European Molecular Computing Consortium.

He has been a member of the program committees for practically all major conferences in theoretical computer science in Europe.

Professor Rozenberg is currently involved in a number of externally funded research projects on both national and international levels. His current research interests are:

- DNA computing,
- theory of concurrent systems, in particular, theory of Petri nets, theory of transition systems, and theory of traces,
- theory of graph transformations,
- formal language and automata theory,
- mathematical structures useful in computer science, in particular, theory of 2-structures,
- and, Computer Supported Cooperative Work.

Apart from his many academic achievements, he is a devoted husband and proud father, and even an accomplished magician. In fact, he claims that magic is his career and computer science his hobby. His wife and traveling companion, Maja, (who is also a medical doctor) has taken good care of him in recent years as he faces challenges with his health. He could not imagine his life without her. When it comes to heroes, Professor Rozenberg has not found his in movies or books, or even among scientists or magicians, rather his hero is his mother, to whom he owes a great deal.

Today, he looks back on a very satisfying life. In his own words, "I have a wonderful family, I have written many papers, I have shuffled many decks of cards. Life has been good to me."

ARTO SALOMAA

Born in Turku, Finland, on June 6, 1934, Salomaa received his undergraduate and Master's degrees from the University of Turku. He then did graduate studies at the University of California and at the University of Helsinki before receiving his Ph. D. from the University of Turku in 1960.

Salomaa has worked in universities in Finland since 1957, primarily as a Professor of Mathematics at the University of Turku and as a Research Professor and Academy Professor at the Academy of Finland. He has also had extended stays as a visiting and Adjunct Professor of Computer Science at the University of Western Ontario (his first visit was 1966-68), visiting Professor at the University of Aarhus, and visiting Professor at the University of Waterloo; as well as shorter visits to 150 universities in Europe, North America and Asia.

More than ten books have been authored by Salomaa, including *Jewels of Formal Language Theory* (1981), *DNA Computing: new computing paradigms* (1998), and the classic formal language theory book *Formal Languages* (1973) which is widely used as a text and reference book all over the world.

His primary research has been directed towards language theory (his recent interests have been in splicing and replication mechanisms motivated by DNA-studies, patterns and pattern languages, shuffle operations, and complexity and primitivity issues associated to the Post Correspondence Problem), cryptography (particularly voting protocols and cooperative hashing) and algorithmic information theory.

Salomaa has received a number of awards and honours, including the degree of doctor honoris causa from Åbo Akademi, University of Oulu, University of Szeged, University of Bucharest, TU Magdeburg, the State University of Latvia, and TU Graz. He was awarded the yearly prize of Suomen Kulttuurirahasto (The Foundation for the Finnish Culture) in 1986, the Magnus Ehrnrooth prize of the Societas Scientiarum Fennica in 1991, was named 'Professor of the Year' in Finland in 1994, was listed among the top research leaders (top ten or top twelve) in Finland by the Academy of Finland and the University Council of Finland from 1993–97, and received the yearly prize of the Nokia Foundation in 1998. Several books and journals have also been dedicated to him, the most recent being 'Jewels are Forever' (Springer-Verlag, 1999).

He was a long standing editor of the Journal of Symbolic Logic, and of Elektronische Informationsverarbeitung und Kybernetik (now the Journal of Information Processing and Cybernetics); and is currently an editor for Information and Control (New name: Information and Computation), Acta Informatica, Theoretical Computer Science, Discrete Applied Mathematics, Fundamenta Informaticae, Acta Cybernetica, Journal of Computer Science and Technology (Beijing), Journal of Universal Computer Science, Grammars, International Journal for the Foundations of Computer Science (managing editor), and Mathematica Japonica (honourary editor). He is also editor of the Springer-Verlag series of books 'EATCS Monographs in Theoretical Computer Science'.

Arto Salomaa is also a Member of the Academy of Sciences of Finland, of the Swedish Academy of Sciences of Finland, of the Academia Europaea, of the Hungarian Academy of Sciences, and of the European Molecular Computing Consortium. He has been on the program committees of numerous conferences, for instance, 18 ICALP's. Finally, he has authored more than 400 scientific publications in major journals.

Saunas are very important for Salomaa: he believes that the heat opens the blood vessels in the brain, and that this helps stimulates thought. Furthermore, the complexity of an open problem can be expressed in terms of sauna time - more simple problems may only require one session in the sauna, whereas a more complex one may require two or three visits.